EPIC RIVALRY

THE INSIDE STORY OF THE SOVIET AND AMERICAN SPACE RACE

Von Hardesty and Gene Eisman

NATIONAL GEOGRAPHIC

WASHINGTON, D.C.

OPPOSITE: *Near and yet so far: A full moon rises over Cape Kennedy during Apollo 11 launch preparations, July 1969.*

ISBN: 978-1-4262-0119-6

Library of Congress Cataloging-in-Publication Data

Hardesty, Von, 1939-
  Epic rivalry : the inside story of the Soviet and American space race / by Von Hardesty and Gene Eisman ; foreword by Sergei Khruschev.
    p. cm.
  Includes bibliographical references and index.
  ISBN 978-1-4262-0119-6 (hardcover : alk. paper)
  1. Space race. 2. Astronautics--United States--History--20th century. 3. Astronautics--Soviet Union--History--20th century.
I. Eisman, Gene. II. Khrushchev, Sergei. III. Title.
  TL788.5.H37 2007
  629.409'046--dc22

                        2007017393

Founded in 1888, the National Geographic Society is one of the largest nonprofit scientific and educational organizations in the world. It reaches more than 285 million people worldwide each month through its official journal, NATIONAL GEOGRAPHIC, and its four other magazines; the National Geographic Channel; television documentaries; radio programs; films; books; videos and DVDs; maps; and interactive media. National Geographic has funded more than 8,000 scientific research projects and supports an education program combating geographic illiteracy.

For more information, please call
1-800-NGS LINE (647-5463)
or write to the following address:

NATIONAL GEOGRAPHIC SOCIETY
1145 17th Street N.W.
Washington, DC 20036-4688 U.S.A.

Visit us online at www.nationalgeographic.com/books

For information about special discounts for bulk purchases, please contact National Geographic Books special sales: ngspecsales@ngs.org.

Printed in U.S.A.

Interior Design: Melissa Farris

CONTENTS

HOW ROCKETS LEARNED TO FLY

Foreword by Dr. Sergei Khrushchev

I had just turned 22 when, on October 4, 1957, the first man-made satellite was launched from the Baikonur launch site, which at that time still had the same name as the closest train station—Tyuratam. I cannot say that I was surprised, since for the previous 10 years I had been an avid reader of science fiction that portrayed man's adventures in outer space. The stories were published en masse in the Soviet Union. For college students and schoolchildren of that time, the idea of a flight beyond Earth's atmosphere would be an extraordinary achievement but not a feat beyond the realm of possibility. We were more perplexed that the satellite had not been launched sooner.

Humankind had been preparing for a spaceflight for several decades. In Russia, Konstantin Tsiolkovsky had spelled out the basic formulas for a spaceflight; in the United States, Robert Goddard had launched his model rockets; and in Germany, Hermann Oberth had tested rocket prototypes. Any one of these countries could have been the first, but the Soviet Union was destined to pave the way.

OPPOSITE: *Sergei Khrushchev (left) with his father, Nikita, seven years after the father's 1964 removal from power.*

Why? There were some very pragmatic reasons for the Soviet triumph. After World War II, only two competing powers were left: the United States and the Soviet Union. In the postwar era, nobody would have been surprised if these two countries had gone to war. Both sides were certainly preparing for it. In the 1950s, the United States enjoyed indisputable superiority, having ringed the Soviet Union with air bases. American strategic bombers were capable of turning any Soviet city into another Hiroshima or Dresden. The leadership of the Soviet Union feared a nuclear catastrophe and did everything possible to avoid it.

But how to avert such a catastrophe? Soviet bombers were incapable of striking any target in the United States—even theoretically. The Kremlin had no option but to call for peace—and, feverishly, to seek a military option to oppose America's offensive and destructive air power. The situation became particularly nerve-racking in July 1956 after U-2 spy planes started flying regularly over Soviet territory, even missions over Moscow and Leningrad. No one had any doubts: The U-2s were identifying targets just as the Germans had done in the spring of 1941 before they attacked the Soviet Union.

Soviet scientists proposed that intercontinental ballistic missiles could ensure parity with America. But no one could figure out how they would fly the distance of 5,000 miles to reach targets on the other side of the world. Nevertheless, Soviet leaders, including Nikita Khrushchev, regarded rockets as a way to provide a measure of national security.

Renowned aircraft designer Semyon Lavochkin started developing an intercontinental missile—a flying bomb, as it was called back then—a design with a ramjet developed by Mikhail M. Bondaryuk. Meanwhile, Sergei Korolev, then a relatively unknown figure, promised to solve the problem by means of a ballistic missile. Born in 1907, Korolev had devoted his life to aeronautics. As a youth, he had built gliders, then boost-glide aircraft, and in the mid-1930s he actually succeeded in building rockets. Korolev also ended up in the gulag—where by all accounts he narrowly escaped death. His imprisonment could be traced to his rocket research and the fact that Marshal Mikhail Tukhachevsky, a big proponent of rocket building, had fallen out of favor with Stalin. But Korolev did not perish in the camps; he was miraculously saved. At the time of the war, Korolev was in prison, freezing in temperatures of 40°F below zero. He was returned to the "mainland" and was assigned to build rocket boosters for military bombers. In this prison workshop, Korolev drew his blueprints, while two hefty guards watched over him and his work.

The Soviet views started to change in 1945, when it came to light that the Germans were far ahead of the Allies in rocket technology. Stalin ordered the release of all rocket specialists whom he had not yet executed, assigned them military ranks of colonel or captain, and sent them off to Germany to seek out remnants of the advanced German rocketry. Given the rank of lieutenant colonel, Korolev was assigned to conduct research on the German V-2—far from a lofty assignment.

Sergei Korolev, a born leader, did not stay long in a third-rate position; a few years later, he had already become Chief Rocket Designer. In the grand scheme of the military, this was not a great position, but within the Arms Ministry it was quite significant. Korolev designed his R-1, R-2, and R-3 rockets based on the V-2. He also eclipsed his opponents. Under the pretext of maintaining secrecy, Korolev barred his most dangerous rivals—the "captive" German rocket specialists under the leadership of Helmut Gröttrup—from all practical work and sent them to the "comfortable" island of Gorodomlya in Lake Seliger, halfway between Moscow and Leningrad. From there the well-disciplined Germans kept sending quite promising proposals to Moscow. They far surpassed the work done by Korolev himself, but the reports ended up on Korolev's desk for review, after which they ended up in the archives. Most of the imprisoned Germans, including the rocket scientists, had been sent back home to Germany by the mid-1950s. This agreement coincided with German Chancellor Konrad Adenauer's visit to Moscow in September 1955. Korolev finally sighed with relief.

It was a little more complicated to get rid of another Soviet "German" —Mikhail Yangel—a talented engineer who was Korolev's immediate boss during the first stage of his work. However, Korolev managed to send him to Dnepropetrovsk to "ensure" the manufacture of his rockets at the production line facility. Thus, Korolev was left alone, and his finest hour arrived.

The development of an intercontinental R-7 missile got under way in 1954. Following the first hydrogen bomb test in the Soviet Union in August

1953, military customers insisted on equipping their "principal weapon" with a three-megaton thermonuclear charge. Andrei Sakharov conducted an analysis, and he reported that the charge would weigh about five tons. Overall, it was projected that the warhead would weigh 6.5 tons, including the charge itself, heat-resistant coating, plus everything else. As it turned out, Sakharov had made a mistake, and the thermonuclear charge turned out to be lighter. However, his mistake would predetermine the success of Soviet astronautics for many years to come.

Korolev had shed himself of his rivals, but he was in need of talented specialists—something he understood very well. Then he and everyone around him got incredibly lucky when Mikhail Tikhonravov appeared on the horizon. Seven years older than Korolev, Tikhonravov had taught him how to build rockets in the 1930s. Unlike Korolev, Tikhonravov had not been arrested for some reason. He was drafted into the army, where he served all these years quietly and inconspicuously, collected butterflies, and dreamed about a rocket flying into space.

The problem to be solved was putting a spacecraft into orbit around Earth. According to the theorem proven by Tsiolkovsky, a one-stage rocket would not work. Tsiolkovsky reasoned that only a multi-stage rocket could deliver a satellite into the Earth's orbit. But the dreamer-mathematician did not focus on how to make that rocket fly. The solution to that problem fell to Tikhonravov. The essential requirement boiled down to the rocket-engine firing. On the ground, due to gravity, all the fuel components and the oxidizer dutifully flowed from the tanks down the tubes to the aft part of the rocket and the engine. But the engines of the second and the subsequent stages had to be fired in zero gravity. Where would the fluids flow and would they flow at all? Nobody knew the answer to this question, and consequently, no one could guarantee that the second-stage engine would ignite. There was no way to verify it, either. The task seemed to be insoluble, but Tikhonravov found an answer. He came up with the idea of bundling several stage-one rockets into one packet. Their engines were fired simultaneously, while still on the ground, and once in flight, as the fuel was depleted, some of the rockets were jettisoned, and the remaining ones had enough power to deliver the satellite into orbit. However, this proposition was still not enough to solve the entire problem. It proved equally difficult to work out a solution "in metal." It required intelligence and a lot of energy. Tikhonravov, not a very practical person and definitely lacking in organizational skills, did not take this aspect into account. He published an article in a scientific publication, made a presentation at a conference, and sent a report to the military authorities.

Both Korolev and Tikhonravov worked in the Moscow suburbs, Korolev in Podlipki and Tikhonravov in Bolshevo. They met by chance, had a drink, shared stories of the old times, and talked about the future. Tikhonravov talked about the rocket "packet," and Korolev understood immediately. It was exactly what he needed, and he invited Tikhonravov to work in his design bureau. An armchair scientist, Tikhonravov did not have the slightest chance of implementing his ideas without Korolev. By the same token, the brilliant project manager Korolev needed a visionary like Tikhonravov. Together they constituted the "critical mass" that shook the whole world.

During the initial stages of the rocket's development no one mentioned a satellite. Not until mid-1955, at the time when the R-7 rocket was being developed, did Tikhonravov gingerly bring up the subject. For him, the rocket had little appeal without a satellite. Korolev had a satellite in mind, too, but he thought that the rocket was more important since it was the top priority for the government. In 1955, a man-made satellite in orbit around Earth was viewed as a toy, a "scientific amusement" with no practical application. Korolev knew that persuading the leaders of the Soviet state to spend funds on "nothing" would be difficult. But they might be interested in setting a new world record—and at practically no extra cost. Accordingly, a plan took shape in Korolev's mind.

In January 1956, Korolev, then a rank-and-file chief designer, did not have direct access to Khrushchev. Nevertheless, he used all his organizational skills to push through a government resolution that provided for "the launch of a man-made satellite using the R-7 intercontinental missile in the interests of the U.S.S.R. Academy of Sciences." But this resolution did not mean that the project would necessarily get the green light. The Soviet bureaucracy moved according to its own understanding of what was really important and what was not. A critical breakthrough came on February 27,

1956, when my father, Nikita Khrushchev, made a visit to Podlipki with several other Soviet leaders. He wished to meet Korolev in person, to look into his eyes, and to convince himself of the reality of the ballistic rocket to which he would entrust the future of the country's safety. My father took me with him. I was a student at the Moscow Electric Power Institute, and he believed that it would be useful for me, a future engineer specializing in automated control systems, to be well informed about recent technological achievements. That day I shook hands with Korolev, a person who would soon be known as a great man. I had no idea then that barely two years later I would become a rocket engineer and fate would bring me together with yet another great man, Vladimir Chelomei, and that I would in some way participate in the competition between these two icons. In my book *Nikita Khrushchev and the Creation of a Superpower,* I wrote about the relationship between these two great men and their interaction with Khrushchev, Soviet officialdom, and the military.

On that pivotal day of February 27, after a conversation about important military matters, Korolev led my father to a special screened-off area in a large hangar. In the discussion that followed, he showed Khrushchev a modest poster with the design for placing a satellite in Earth orbit. Korolev asked for support, arguing that the satellite would not take one minute away from the "main project," which was the development of an intercontinental missile. My father, an intellectually curious person by nature, promised his support; he also wanted to know what was out there beyond the Earth's atmosphere. He reiterated that the country's safety was priority number one. Everyone present in the hangar that day witnessed Korolev telling Khrushchev about the satellite and Khrushchev agreeably nodding his head. The words that were exchanged were not that important; what was really important was that Korolev secured the personal support of the number one person in the country, which in Russia means much more than any government resolution.

Now the future depended on Korolev himself. He needed to overcome serious obstacles to surpass the Americans. He started with the "academics." They were working slowly, stuffing the satellite with as many devices and sensors as possible—tests that were important for science, but not for Korolev. At their rate the project could go on for months and months. Korolev ordered the removal of all the scientific add-ons, making the satellite as simple as possible. The result is known to the entire world as the hollow, polished ball with four whiskers of antennas—an ingenious solution by a brilliant manager. Given the secrecy of the time, he worked as best he could. He also launched a publicity campaign. Newspapers and magazines carried articles here and there about an upcoming launch of a satellite, but gave no specific information on the date. The magazine *Radio* even published the satellite's transmitter frequencies. The world paid little attention to those publications.

All these preparations, though, were worthless without a rocket. It was necessary to teach the rocket how to fly, and the rocket was in no hurry to learn. Starting in May 1957, one failure followed another. Then the Americans successfully tested their Jupiter-C rocket in August 1957. Korolev was crestfallen. He had no doubts that a satellite would follow the first successful launch—at least that was his intention. But the Americans were moving slowly. Korolev suspected that they were plotting something. But what was it?

Finally on August 24, the "Semyorka," as the R-7 intercontinental missile was nicknamed, was launched and reached distant Kamchatka. Another successful launch followed in September. Now it was the satellite's turn. Korolev was in a great hurry. The International Geophysical Year (IGY) coordinating committee scheduled its meeting for early October, and the agenda included a presentation by American scientists on their projected Vanguard satellite. Korolev put himself in their place, and he had no doubts that the Americans would prefer using the Jupiter-C, a rocket already successfully tested, to launch a satellite into orbit. Korolev then ordered his staff to work day and night in order to launch his satellite ahead of the Americans, perhaps one day before the IGY meeting.

By pure chance, on October 4, 1957, I happened to be with my father at the Mariinsky Palace in Kiev, the capital of Ukraine. Korolev called to report a successful launch. My father and I turned on the radio to listen to Sputnik beeping as it flew over Europe. Neither Khrushchev nor Korolev—and I even less—realized the immensity of what was happening during those hours. The next day *Pravda* and other Soviet newspapers published a two-column standard report on their front page from the Telegraph Agency of the Soviet Union (TASS): "About the Launch of

Earth's Artificial Satellite in the Soviet Union, in accordance with the International Geophysical Year Program." Reports of similar achievements, such as the first nuclear power plant in operation, the first jet aircraft flight, or a new high-altitude record set by the new MiG jet fighter were standard fare. The article about the first satellite was no different.

Moscow suddenly realized a day later that the satellite had caused quite a furor all over the world, especially in the United States, and the Russian word "sputnik" for "satellite" soon entered the languages of all nations. Ironically, it was the American press, not the Soviet press, which gave the Sputnik launch such immense coverage, allowing it to become one of the most powerful weapons of propaganda the Soviet Union had.

After Sputnik, Korolev could call Khrushchev any time, knowing that as the Soviet leader he would answer the phone, listen to him, and give the necessary orders no minister would dare ignore. Thus, Korolev gained the power necessary to implement his other ideas. However, Korolev's name was not widely known at the time. He became world famous as the "Chief Designer" spelled with capital letters, and he humbly signed his articles in *Pravda* merely as "Engineer Sergeyev." With such anonymity, Soviet counter-intelligence would force the CIA to spend its resources to uncover "non-secret" secrets, to prevent them from uncovering the real secrets for as long as possible.

Meanwhile, Korolev was in a hurry to solidify his success. After returning to Moscow, my father called Korolev to congratulate him. During that call, Korolev suggested that a new more complex satellite be launched as early as a month later, by November 7, the anniversary of the October Revolution. Naturally, my father did not object. That left him very little time. Korolev, an excellent psychologist, explained to his people that they were working on a project assigned at the highest levels and that he would report to Khrushchev every evening on the daily progress. His ruse worked. Soviet technicians worked day and night to meet the goal. The second satellite was launched on schedule as planned by Korolev.

My pragmatic father did not like to "put all his eggs in one basket," particularly in light of the fact that the military rocket R-7 project did not yield a useful weapon and resulted in large expenditures for each launch pad built. Meanwhile, Mikhail Yangel, who had been "exiled" by Korolev to Dnepropetrovsk, gradually grew stronger and started designing his own rockets—initially medium-range types. When he proposed building the relatively "inexpensive" R-16 intercontinental missile using high-boiling-point components, dimethyl-hydrazine and nitrogen-tetroxide instead of oxygen-kerosene, and navigating by using a gyroscope instead of a radio beam, my father reacted to the idea favorably. He signed the government's decree authorizing the R-16 project.

In Russia, however, a formal document does not mean that everything is ready to go. My father did not risk making a final decision without consulting Korolev first. He invited Korolev to the Kremlin, but they did not have an amicable meeting. As soon as Korolev heard the name Yangel, he kicked up a fuss and denounced the project as a reckless scheme. He claimed that engines working with high-temperature components were not technologically feasible for intercontinental ranges. My father became upset at this news. The nation had gained a satellite but remained defenseless in the face of an American missile attack. That sense of hopelessness, and not a lack of trust in Korolev, prompted my father to double-check with "Korolev's" engine specialist Valentin Glushko. Much to my father's surprise, Glushko held the opposite opinion and promised to make Yangel's much-needed intercontinental missile engines on time.

Korolev now faced a dangerous rival, and he did not tolerate competition. He focused his anger on Glushko, calling him a snake in the grass, saying that he would never shake his hand, and he would never work with him again. However, after calming down a bit, Korolev tried to straighten out the situation by gaining control of Yangel's rocket, but this transfer of power was not granted. Khrushchev not only disagreed with Korolev, but he also started having doubts about Korolev's objectivity. That's how Korolev lost both his indisputable authority and his invulnerability. Korolev was not entirely upset. The space programs were still under his control. But the military was already gossiping about Korolev. In time, the military placed its money on Yangel.

Meanwhile, space mania became more fashionable. Korolev dragooned aviation designers against their will into space projects; they had never even dreamed about flights beyond Earth's atmosphere. My boss, Vladimir Chelomei, who had been involved exclusively in submarine-launched missiles, was among them. Now we were drawing winged spacecraft that looked somewhat like the contemporary space shuttle. They looked appealing to government officials when presented on flip charts with purple, dark-green, or charcoal-black, outer-space-like backgrounds. Chelomei wanted to make them as real as possible, and he exasperated his deputy, Mikhail Lifshitz, by claiming that he just could not get the color of space right. And what was the "right" color of space? No one had seen it yet.

Korolev was getting ready to send a man into space, while Chelomei was drawing his posters. Then came April 12, 1961. Radio reports announced Yuri Gagarin's flight on the orbital spacecraft Vostok. He was a lieutenant at liftoff and promoted to major by the time he had landed. Khrushchev had a confrontation with Minister of Defense Marshal Rodion Malinovsky over that promotion. He believed that it was inappropriate even for a cosmonaut to jump rank, but he signed the order after grumbling a bit.

My father greeted Gagarin at Moscow's Vnukovo airport. While Gagarin was marching up the red carpet, a rubber garter holding his sock up came undone and kept striking his leg and causing pain. Everything else went just fine. Gagarin's triumphal procession moved along Moscow's streets with the cosmonaut standing in the open limousine, while my father sat humbly in the back seat. The entire city cheered. People were hanging out the windows; there were crowds on the rooftops. A meeting in Red Square followed, then a reception in the Kremlin, and finally an awards presentation. The exultation at Gagarin's celebration could only be compared to the victorious rapture of May 9, 1945, when Germany formally signed its terms of surrender.

Korolev was humble at the reception in the Kremlin, knowing that everyone with the need-to-know understood that it was *his* celebration. But neither he nor the others knew that Korolev had reached his acme that day. He did create the R-7 rocket, which would serve mankind for a half-century, but Korolev would never provide a successor.

Soon after Sputnik, Korolev ordered his subordinates to start designing a new, more powerful rocket, the N-1. This huge rocket would be capable of delivering a record 30-ton payload into orbit. But Korolev had no clear understanding of what one could do with those 30 tons. Worse, the N-1 was designed by "artisans," who were skilled professional engineers, but lacked any "divine spark." Tikhonravov was not interested in the rockets anymore; he was working in the cosmos area again to his heart's delight and became the head of the satellites department at Korolev's design bureau. Korolev never found another talented rocket specialist of the same caliber. As a result, all his energy and talent as an organizer and manager "slipped through the sand." The rocket design took shape, but it was mundane and bland; moreover, no super, high-powered engines were earmarked for it. Glushko was now working with Yangel and Chelomei—with anyone but Korolev—and he remarked sarcastically that his "engines would shoot any piece of metal into space." However, this particular "piece of metal" would be a rocket without his engines. Korolev gave up on Glushko, too, and said publicly that there are no indispensable specialists; one only has to look harder.

In May 1961, President John Kennedy challenged the Soviet Union to a competition to land the first man on the moon. The Semyorka was not suitable for this new kind of mission. The N-1 project soon stalled, and Khrushchev was in no hurry to accept the challenge. It was one thing to set records using readily available R-7s and not spending a lot of money. It was entirely different to undertake another high-stakes project without any practical benefits. It was his understanding that any moon race would be costly, and he felt the outcome did not justify the expense. At the June 1961 meeting in Vienna, Austria, Kennedy tried to sound Khrushchev out and get a feeling for whether the two nations could undertake a joint moon project.

On his side, Kennedy did not want to risk the Soviets being first again. He felt that it would be better to share success with a rival than to be left alone in ignominious defeat. But Khrushchev did not respond to Kennedy's overtures. Let the Americans waste their dollars; he would find better use for his rubles. Housing projects and food production, not space, were the priorities for Khrushchev.

Meanwhile, Korolev was champing at the bit and inundated Khrushchev with memorandum after memorandum, trying to prove that his N-1 could be the first to land a man on the moon. It would cost him very little since, unlike the Americans, he already had a head start. Khrushchev requested a cost estimate. Korolev prepared several estimates, since it was not easy to determine the cost of things in the Soviet economy. Finally, in February 1962, Korolev managed to get permission to start work. In August 1964, he literally wrested consent from Khrushchev for the launch of a full-scale project to design a moon-landing spacecraft. However, the approval came with a strict provision that the project must stay within budget.

In October 1964, Khrushchev was removed from power. In the period that followed, nobody monitored whether Korolev was staying within budget. Nobody even asked Korolev about it. Brezhnev showed little interest in such trivial affairs, although the idea of jumping ahead of the Americans appealed to his vanity. Consequently, Korolev was given a green light. The race to the moon took off at full speed. However, a third participant suddenly appeared on the scene—Vladimir Chelomei. He and Glushko proposed their own moon project based on the UR-700 launch vehicle. Korolev then focused his energy on what he did best—the elimination of his rivals. My friend Vladimir Modestov, who was Chelomei's deputy on guidance matters, used to joke bitterly that these two, Korolev and Chelomei, would rather see the Americans land first on the moon than to see the other succeed. Toward the end of 1965, Korolev finally succeeded in his campaign against Chelomei and Glushko. He kicked both of them off the moon project. Yet Korolev was left with a Pyrrhic victory.

By his own actions, Korolev had cleared the way for Wernher von Braun to beat him to the moon. Korolev had lost the competition even before it had started. Having ordered the design of the N-1 for a future mission to the moon, Korolev—as it turned out—had approached the problem backward. When calculating the weight needed at liftoff, one usually starts with the finish line: identify the spacecraft's mass on the moon, then in the moon's orbit, then in Earth's orbit, and finally on the ground, with a 10-15 percent margin of error. As a result of this analysis, the Americans figured out what they would need to deliver the 129-ton Apollo spacecraft on a 2,900-ton Saturn V launch vehicle into Earth orbit. Taking into account the level of Soviet technology, especially in the area of electronics, the spacecraft would need to be at least 145 tons, which translated into a 4,500-ton UR-700 rocket on the launch pad at Baikonur. I myself, working for Chelomei, participated in this analysis.

Korolev was firmly holding on to his N-1, yet from the very beginning it was "Trishka's caftan" (a Russian expression literally meaning "poorly tailored coat"). In 1964, after multiple alterations, it was able to lift not only a 30-ton payload into orbit but eventually a 70-ton payload. That was a lot, but insufficient for a mission to the moon. Yet again, Korolev ordered that everything be redesigned to increase the spacecraft mass to 90-plus tons. Additional engines had to be added to the first stage. The engines had their own problems, since Korolev had gotten rid of "his engine expert" Glushko. He did find a replacement for him, hiring Nikolai Kuznetsov, but Kuznetsov was not a rocket designer. He was a turbojet aviation engine designer—a talented specialist but from a different field. He had to learn how to build rocket engines, but there was no time for that.

What happened next is well known: On July 21, 1969, Neil Armstrong set foot on the moon and uttered the phrase: "One small step for man, one giant leap for mankind." There was a triumphal return to Earth. Wernher von Braun won the race to the moon. However, Korolev was still the trailblazer. Just as Christopher Columbus discovered America, Sergei Korolev had laid the groundwork for mankind's journey into space. No one following in his footsteps will ever catch up with him.

In this book, Von Hardesty and Gene Eisman, in an entertaining and highly detailed way, narrate the beginning of the space age. They show how Sergei Korolev in the Soviet Union and Wernher von Braun in America were able to lead humankind beyond the boundaries of planet Earth through their perseverance and their ability to mobilize and unite huge collectives around them and to persuade the authorities to do what they considered important. The authors find dramatic and sometimes amusing episodes from the lives of heroes who didn't consider themselves heroes at all; to them, they were just doing their jobs. It was to this end that they devoted their entire lives.

Now please read and enjoy.

—*Dr. Sergei Khrushchev*
March 22, 2007

INTRODUCTION

The Epic Rivalry

Threspace race may be defined broadly as the 12-year competition between the United States and the Soviet Union for dominance in the new frontier of space. A memorable episode in the Cold War era, this superpower rivalry—at once spirited, high-risk, and costly—developed roughly between the launch of Sputnik in October 1957 and the Apollo 11 lunar landing in July 1969. This book tells the story from both sides, using the rich body of English and Russian language sources now available to researchers. The narrative has a dual approach: to reconstruct the parallel universes of the American and Russian space programs—and then to identify how these separate worlds interacted in necessary and fateful ways. The authors have selected those key events, personalities, and technologies that shaped the course of the space race. *Epic Rivalry* appears in the 50th anniversary year of the launch of Sputnik, an occasion to reappraise how the vision and reality of space exploration shaped the modern world.

OPPOSITE: *A Russian illustration celebrates the 1957 launches of Sputnik 1 and 2, the first artificial satellites.*

Looking back, the Apollo 11 mission gave expression to the ancient urge by humans to explore distant worlds. This latter-day Homeric journey to the moon was televised directly to a global audience still caught firmly in the grasp of Earth's gravity. Countless earthbound observers watched the dramatic moon landing. The lunar trek represented an engineering feat without parallel. When Neil Armstrong took his first step onto the lunar surface, he correctly described that moment as a "giant leap for mankind." For space visionaries, Apollo 11 pointed to the possibility of future journeys to Mars and even more distant locales in the solar system. Yet this seminal moment occurred within a distinct historical context—that of the Cold War rivalry between the United States and the Soviet Union.

The story of the space race survives in human memory in peculiar ways. Now only a minority remember the tremendous impact the 1957 launch of Sputnik made on Americans, in particular their cherished sense of superiority in the realm of modern technology. Looking back, many Russians today remember the first orbital mission of Yuri Gagarin as the singular milestone of the space age; Americans, by contrast, often regard the Apollo 11 mission with Neil Armstrong stepping onto the lunar surface as the pivotal moment. Also, most Americans know there was a space race, and the United States placed the first humans on the moon in 1969, but they may not know that the Soviet Union mounted a serious manned lunar program of it own, in direct competition to the Apollo program. Yet while America's huge Saturn V rocket enjoyed a 100 percent success record on its lunar missions, the Soviet equivalent, the N-1—in four attempts—never got very far off its launch pad before exploding into a fiery conflagration.

"I believe," President John F. Kennedy stated on May 25, 1961, "the nation should commit itself to achieving the goal, before this decade is out, of landing a man on the moon and returning him safely to Earth. No single space project in this period will be more impressive to mankind or more important for the long-range exploration of space." Kennedy's words, spoken in the immediate aftermath of Yuri Gagarin's orbital flight, focused the emerging American space program on the lofty—and exceedingly difficult—goal of flying to the moon. The untested new president spoke boldly at a time when the Soviets appeared to hold a decisive edge in space technology. Kennedy's utterance keynoted a decade of space rivalry as each side worked to perfect the rockets and techniques to reach the moon—still a distant celestial body.

This race became a fascinating study in contrasts. Both nations used their existing military technology to help fashion their space programs. While the American space program—except for its military aspects—remained open and dependent on public support, the Soviets operated under a shroud of secrecy, studiously concealing from view their specific goals in space for the near and long term, even declining to reveal the names of its chief space leaders.

This competition prompted each nation to allocate vast human and economic resources for space exploration. In the end, though, more than just budgetary allocations determined the victor. Ultimate success rested on the technological, industrial, and organizational capacities to sustain a coherent space program. Each side had to integrate baseline military programs with new space priorities.

The story told in *Epic Rivalry* predates the advent of modern rocketry. Indeed, the human interest in space travel and visiting the moon has its origins in ancient mythology. In more recent times, visionaries gave expression to the dream of a lunar trek in literature, art, music, and film. For these earthbound dreamers, the moon—with its powerful impact on the Earth's tides and its hidden far side—remained beyond reach, notwithstanding its apparent close proximity as a celestial object. With the start of the 20th century, the gap between science fiction and science fact narrowed through the theoretical work of a group of space pioneers—Konstantin Tsiolkovsky, Hermann Oberth, and Robert Goddard. These men grappled with the actual technical requirements for space travel.

The advent of liquid-propellant rockets in the 20th century paved the way for the space age. This pioneering form of propulsion was volatile and dangerous, but it offered the best option to overcome gravity and catapult humans into outer space. This evolving technology reached maturity with the German V-2 rocket, first tested successfully at Peenemünde in October 1942. On that fateful launch, the V-2 took a trajectory through the upper atmosphere to the edge of outer space. As a weapon, the V-2 failed to reverse the course of the war, but it did represent a radical new technology. As the war ended, the United States and the Soviet Union raced to capture the surviving V-2 rockets. The wartime spoils of German technology became the basis for a new generation of rockets, prized initially as a delivery vehicle for nuclear weapons. Starting in the late 1950s, both superpowers moved to adapt their evolving rocket technology for space exploration.

This book examines the historic role of the key political leaders who shaped the course of space exploration. Nikita Khrushchev, who consolidated his power in the Soviet Union by the mid-1950s, took a keen interest in space activities, seeing clearly the propaganda value of a series of space "firsts" with the launch of satellites, space probes, and manned orbital missions. The United States embraced the space age with a more cautious posture, at least initially. In the 1950s, Dwight Eisenhower was hesitant to pursue expensive space programs, often preferring more narrowly defined scientific goals rather than manned missions. Eisenhower also feared an unbridled pursuit of technology for military or civilian ends. John F. Kennedy, who assumed office in 1961, presided over a shift in American space policy. While not personally interested in space, he came to recognize its importance in the Cold War context. In the immediate aftermath of the Soviet Union's first manned orbital flight, Kennedy decided to commit the United States to the goal of a lunar mission, then considered by many as a most fanciful idea. His name would eventually adorn the spaceport in Florida. Lyndon Johnson had been alarmed at the Sputnik "surprise," and he advocated a proactive space program for the United States. Richard Nixon affirmed the priorities of his predecessors, took great delight in the triumphs of the Apollo program, but eventually cooled to the idea of an expansive NASA program with all of its budgetary burdens.

Success in either the American or Soviet space programs rested on the innovative work of talented scientists, engineers, and managers.

Wernher von Braun came to the United States in the post-war period with an experienced group of German rocket technicians. This core group did much to advance the American rocket program. The creation of NASA in 1958 consolidated America's space-related assets. Later, James Webb, assisted by an able group of administrators, oversaw the efficient management of the NASA space program—one committed to the fulfillment of the Kennedy mandate for a lunar landing. On the Soviet side, the space program would be directed by Sergei Korolev, the mysterious "Chief Designer." Korolev, a former political prisoner, became a remarkable and singular administrative force shaping the Soviet space program—often against formidable odds and opposition from the military. At the remote spaceport of Baikonur, he, too, worked with a group of highly motivated, if often contentious, group of designers and engineers. Both superpowers pursued a space program in lockstep with missile research and development.

The pursuit of a competitive space program required the recasting of national priorities. In the United States public enthusiasm for space and the financial largess of Congress became essential for success. Throughout the time frame of the space race, NASA managed to sustain broad, often uncritical, support for its goals. This situation would change in the post-Apollo years. Contrary voices gained a new momentum and a larger audience. Many sacrosanct NASA programs now came under increased scrutiny. There were questions raised about cost overruns, misplaced priorities, and the problematical nature of manned space missions. Putting humans in space, some argued, was complex, costly, and dangerous. Accidents such as the Apollo 1 fire brought unwelcome scrutiny to NASA. The tragedy reminded the public at large that adventure in space travel did not come without risk. In its formative years, though, NASA benefited from a talented and hard-driving management—one willing to take risks. Those within and outside NASA advocating greater stress on scientific research routinely encountered obstacles; when it came to budgetary allocations, manned space activities routinely enjoyed a powerful mandate. NASA was always in motion, purpose-oriented, and evolving as an organization—and rarely without a loud chorus of critics.

Essential to any space program—in the United States and the Soviet Union—was the recruitment of space voyagers: the highly skilled, courageous astronauts and cosmonauts. Initially, they were drawn from the realm of military pilots, many of them combat veterans or experienced test pilots. The early pioneers, including Russia's Yuri Gagarin, German Titov, and Aleksei Leonov, mesmerized the world with their pioneering space feats. Their American counterparts, Alan Shepard, Gus Grissom, and John Glenn, displayed equal courage and became instant celebrities. Subsequent generations carried on the momentum of spaceflights in increasingly powerful and sophisticated spacecraft. The Apollo 11 mission with Neil Armstrong, Buzz Aldrin, and Michael Collins brought a decade of heroic work by astronauts to a fitting conclusion.

This book offers a glimpse into one of the most dramatic episodes in the history of the Cold War, the space race between the United States and the Soviet Union. This intense, if brief, interlude of superpower competition helped to shape a new era of space exploration, one which continues unabated into the 21st century.

The production of *Epic Rivalry: The Inside Story of the Soviet and American Space Race* involved the creative contributions of many talented individuals. We wish to acknowledge the overall guidance of our editor at National Geographic, Garrett Brown, who worked with us tirelessly, always providing prompt, wise, and helpful support and advice on all phases of our work. John Paine played a key role as a text editor, offering timely commentary to shape the content and focus of the text. National Geographic illustrations editor Olivier Picard did an excellent job in seeking out vivid and often rare images associated with the space era. The book's cover and interior design benefited from the creative work of Melissa Farris. Michael Gorn, a historian at NASA's Dryden Research Center at Edwards Air Force Base in California, played the dual role of helping to shape the concept of the book and offering technical advice. Dmitry Sobolev, a Russian historian, made substantial contributions to the research for this project. We also owe a debt of gratitude to David West Reynolds, author of *Apollo* and *Kennedy Space Center*, for writing a series of lively and insightful sidebars. We were delighted that Sergei Khrushchev—himself an engineer and one-time participant in the Soviet space program—contributed a foreword and advised us on many historical aspects of the space race. Lastly, we wish to thank our wives, Patricia Hardesty and Charlene Currie, for their unfailing patience and support over the past year while we wrote *Epic Rivalry*.

PROLOGUE

Arrow to the Future

G eneral Walter Dornberger, head of the German army's rocket program, awoke to a brilliant sunrise on October 3, 1942. Posted at Peenemünde, the army's top secret test facility on the Baltic coast, Dornberger greeted the dawn with great anticipation. The weather forecast called for cool temperatures, a blue sky nearly devoid of clouds, unlimited visibility, and minimal winds along this isolated stretch of coastline—the ideal setting for an important, even fateful, rocket launch.

The day had been set aside for the pivotal test of Germany's new secret weapon, the A-4, or V-2 rocket. During the previous summer Dornberger, with his talented technical staff headed by Wernher von Braun, had attempted two launches of the experimental rocket. Both had ended in catastrophic failure. Albert Speer, the Reichminister for armaments and war production, had observed one of the abortive summer launches. He decided to keep an open mind toward the work at Peenemünde, but his support—and ultimately that of Adolf Hitler—remained contingent

OPPOSITE: *A V-2 streaks across the sky during a test launch, Peenemünde, Germany.*

on a successful launch. On this bright October day Dornberger felt an undertow of foreboding—another abortive firing of the V-2 might lead to the cancellation of the rocket program.

Dornberger retained his overall personal optimism, being convinced that the V-2 rocket would ultimately perform well, even flawlessly. At heart, he was a rocket enthusiast, a man who had spent no less than 12 years of pioneering labor in the German rocket program. He knew that his rocket represented enormous potential, a quantum leap in technology that could transform weaponry and offer a vehicle to explore the heavens.

All eyes turned to the imposing V-2 rocket, which had been adorned with a distinctive livery of alternating black and white hues to allow for precise observation and photography of the rocket during its planned trajectory down the Baltic test range. Between two of the fins was a striking image of a girl in black stockings seated on a crescent moon with a rocket in the background. A polite bow to the visionary German movie on futuristic space travel *Frau im Mond* (*The Girl in the Moon*), this peculiar logo bore witness to the fact that the dream of rocketry predated the Nazi regime and Germany's wartime armaments program.

The V-2 was a liquid-fueled rocket, powered by a highly sophisticated engine and controlled by an advanced guidance system. For the German army, then engaged in a titanic struggle against the Allied powers, there was an urgent need for the advanced rocket. Once fitted with a warhead of explosives, the V-2 would become a long-range missile to rain down destruction on Allied targets, particularly London. For Dornberger and von Braun, though, there was another dynamic factor at play. Their prewar vision of rocketry as a means to reach the edge of space and beyond. Such visionary ideas were prudently concealed from their German military overlords.

As the noon hour neared, Dornberger took up a position on the ramparts of a green-painted assembly building adjacent to Test Stand VII. Here the V-2 rocket stood erect with its two umbilical cables attached: one cable to monitor the instruments, the other to provide external power. Beyond the launch pad, Dornberger could look out on the entire Peenemünde test facility—the complex of buildings under camouflage nets, the island of

Greifswalder Oie located across a reed-and-sand-covered promontory, the surrounding pine forests, the Peene River, and the redbrick tower of the Wolgast Cathedral visible on the distant horizon.

Nearby in a concrete shelter, Dr. Walter Thiel, the brilliant designer of the rocket engine, stood ready with his timetable containing the precise launch sequence. To allow for careful monitoring of the rocket at ground level during the risky ignition phase, Thiel and his associates used periscopes mounted on the thick roof of the concrete bunker. Here, too, engineers vigilantly monitored all the complicated internal systems of the rocket through an array of gauges, electrical meters, and other measuring instruments. A final check of all systems, rehearsed carefully in advance, offered reassuring signs that the V-2 rocket was primed and ready. Tension mounted as loud speakers and telephones filled the air with commands, alerting all—engineers, the launch control team, and nearby fire-control units—to the impending firing of the rocket.

With the steering gyroscopes of the V-2 up and running, the loudspeaker began the countdown to the moment of launch.

"X-minus three" sounded across Test Stand VII and beyond.

Nearby, a television camera, as Dornberger remembered, captured the image of the V-2 rocket on a flickering black-and-white screen. The rocket was now ready for its historic flight. A band of condensed moisture at the level of the liquid oxygen tank encircled the cylindrical body of the rocket. Oxygen vapor, escaping from the rocket, was visible at the bottom. All these telltale signs suggested that the critical firing sequence was near at hand. Above, a noontime sun and a clear blue sky beckoned the rocket heavenward.

Numerous engineers and technicians now moved to their assigned positions. From his elevated rampart Dornberger observed the unfolding and carefully scripted ritual. Once the vaporization ceased, he knew that the vent valve on the rocket had been closed, allowing pressure to build in the oxygen tank. The V-2 was seconds away from firing. The last minute of a launch sequence was the most tension-filled, making this brief interlude seem interminable in length, what the launch team called the "Peenemünde minute."

"X-minus one. . . ."

Ignition!

Eight tons of fiery exhaust pushed downward. The two umbilical cables dropped away. The rocket was now on internal power.

Three seconds passed. During this short time frame the thrust increased steadily and dramatically—reaching the requisite 25 tons to lift the 13.5-ton V-2 off its launch pad and into the sky.

Dornberger remembered this singular moment vividly as the rocket leapt from Earth: "Smoke began to darken the picture. Ends of cable, pieces of wood, and bits of grass flew through the air."

As the propulsion engineer pulled the last main lever and released the cables, a turbopump running at 4,000 revolutions per minute with a maximum of 540 horsepower forced 33 gallons of alcohol and oxygen per second into the combustion chamber—a witch's brew of volatile fuel that hurtled the rocket skyward at a phenomenal speed.

The V-2 climbed steadily, leaving in its wake a stream of smoke and dust. Even as the rocket gained altitude, it slowly inclined at a gentle angle, following a predetermined upward path toward the east-northeast.

"It was an unforgettable sight," Dornberger later recalled. "In the full glare of the sunlight the rocket rose higher and higher. The flame darting from the stern was almost as long as the rocket itself. The fiery jet of gas was clear-cut and self-contained. The rocket kept to its course as though running on rails; the first critical moment had passed. . . . The projectile was not spinning; the black and white surface markings facing us did not change. . . . The air was filled with sound like rolling thunder."[1]

The combustion gases escaped from the rocket at a speed of more than 6,500 feet per second, the combustion chamber generating in excess of 650,000 horsepower at the end of the burn time. The rocket, in fact, held a vertical ascent for a mere 4.5 seconds, and then assumed an angle of 50 degrees for its scheduled trajectory down the Baltic test range. Dornberger, with binoculars, nervously watched this critical maneuver, alert to any mishap, fully aware that any loss of power, any breakdown in the sophisticated propulsion and guidance systems, would spell disaster for the rocket program at Peenemünde.

He could hear the noise of the rocket engine mixing with the voice on the loudspeaker monotonously counting the seconds: "14 . . . 15 . . . 16 . . . 17. . . ."

The velocity was now around 650 miles per hour, climbing higher and higher, and then—"sonic velocity"—the V-2 had exceeded the speed of sound!

All these dramatic events passed quickly. The rocket remained stable, under control, with no deviation from the scripted flight path even as it passed through the sound barrier.

Seconds passed. . . . "33 . . . 34 . . . 35. . . ."

The rocket quickly gained speed and altitude. It soon achieved twice the speed of sound (Mach 2) as it climbed to an altitude of six miles.

At 40 seconds . . . another tense moment.

Dornberger and his excited team observed a white cloud appear at the stern of the rocket, prompting some observers to think there had been an explosion. The phenomenon was what they later called "frozen lightning," or the rocket's exhaust of water vapor condensing into a plume.

Reaching the 52-second benchmark . . . the burn was close to shutdown, and still the gleaming V-2 continued to perform flawlessly, still on course, with Peenemünde receding as a distant backdrop.

Now came the critical *Brennschluss* ("end of burning") stage. The command, sent by radio from the ground, closed the fuel valves precisely at the 58-second mark. The rocket engine was shut down.

At this fateful moment, the V-2 was racing across the nearly airless upper atmosphere at a speed of 3,500 miles per hour! Only a thin streak of white condensation clearly marked the remarkable passage of the V-2— now viewed as a faraway speck in the sky and only clearly seen with the aid of binoculars.

"Taking a deep breath," Dornberger recalled, "I put down my binoculars. My heart was beating wildly. The experiment had succeeded. For the first time in the history of the rocket we had sent an automatically controlled rocket missile to the border of the atmosphere at 'Brennschluss' and put it into practically airless space. We had been working 10 years for this day."[2]

Caught up in the excitement of the moment, Dornberger piled into a car with von Braun and dashed to the sand-walled enclosure of Test Stand VII—now empty except for cast-off cables and scattered equipment. Soon they were joined by a stream of technicians and test crew personnel. What followed was a spontaneous outburst of handshakes, cheers, and wild jubilation. The celebration erased all lingering memories of past failures.

The V-2 rocket, traveling at a top speed of 4,440 feet per second, had completed an extraordinary journey of 116 miles across the Baltic Sea east of Peenemünde.[3] During its fall back into the atmosphere, the rocket endured all the stresses of reentry, including the intense heat (1,250 degrees Fahrenheit) and the buffeting that came with the rapid rate of descent (around 2,000 miles per hour). The V-2 nose cone had been fitted with bags of green-colored dye to mark the point of impact in the Baltic Sea. A German aircraft spotted the green-stained impact point an hour later. Dornberger noted that the rocket had struck the water with the energy of 1,400 million foot-pounds, the equivalent of 50 express train engines, each weighing 100 tons and traveling at a speed of 60 miles per hour.[4]

In the aftermath of this dramatic October launch, Dornberger attempted to see beyond the narrow and controlling wartime priorities associated with his rocket. Meeting with his team, he did not speak of any future missile age or even how the rocket might assure a triumph over Germany's formidable enemies. Instead, Dornberger articulated his own expansive sense of the moment, expressing—as he recorded in his memoir—his own breathless "panegyric" on the transcendent historical meaning of his V-2 rocket: "We have invaded space with our rocket and for the first time—mark this well—have used space as a bridge between two points on Earth; we have proved rocket propulsion practicable for space travel. To land, sea, and air may now be added infinite empty space as an area of future intercontinental traffic. . . . This third day October, 1942, is the first of a new era in transportation, that of space travel. . . ."[5] Writing in the postwar years, Speer echoed Dornberger's prescient viewpoint, noting in his memoir, *Inside the Third Reich*, that "for the first time [the V-2] a product of man's inventive mind had grazed the frontiers of space."[6]

The successful launch of the V-2 arguably signaled the advent of a new rocket technology. As a so-called "vengeance weapon," it was deployed too late to reverse the inexorable advance of the Allied powers to victory. In the immediate aftermath of the Allied victory, however, the United States and the Soviet Union engaged in a spirited drive to capture any remnants of the fabled V-2 rocket. This competition set the stage for the epic rivalry that would shape the course of the new space age.

1

THE GREAT TROPHY HUNT

Major Robert Staver of U.S. Army Ordnance arrived in London in early January 1945. His wartime mission aimed to learn more about the V-2, the newly deployed Nazi ballistic missile. This formidable aerial weapon had first appeared in the skies of the British capital the previous autumn. The sudden onslaught of the V-2 evoked fear of a renewed and massive bombing campaign against Britain. The new missile approached its target at a phenomenally high speed with little or no warning. The stark realization that an incoming V-2 could easily overwhelm conventional air defenses prompted grave concerns. What the Nazis called their "vengeance weapon" represented a radical breakthrough in technology, a unique weapon with an unknown potential to wreak havoc in London and to reshape the course of the war.

For Staver, as it turned out, his special mission to London proved to be a perilous journey. The young American officer soon discovered that he was in the crosshairs of the Nazi barrage. On the day Staver reported to

OPPOSITE: *Wernher von Braun briefs high-ranking Nazi officers before a V-2 launch in Peenemünde, 1944.*

his commanding officer, Major Calvin Corey, at 27 Grosvenor Square, he faced the sudden fury of the revolutionary ballistic missile—an incoming V-2 exploded nearby, throwing both men to the floor. When Staver reached a window, he looked out on a scene of devastation: The V-2 had exploded prematurely in the air, raining down debris on the street below.

Shortly after this harrowing incident, Staver moved into his living quarters near Marble Arch. No sooner had he taken up residence in his new apartment than he found himself once again in harm's way. His neighborhood was hit by a V-2 at night, exploding just yards from Staver's bedroom. The impact of the explosion threw him from his bed to the floor. He later observed that the force of the explosion "made the drapes of my window stand straight out from the wall." When another V-2 hit nearby Hyde Park, 62 people were killed. Now, for the first time since the 1940 blitz, Londoners began to look skyward again for any telltale signs of the enemy.

Staver's final encounter with a V-2 came on a leisurely car trip to the Royal Aircraft Establishment at Farnborough: While en route he observed a missile attack on a warehouse just a half mile away, which killed 15 workers and destroyed a large amount of war matériel. For Staver—as with all Londoners that year—V-2 attacks were an unnerving experience: There was no air-raid siren, no effective radar alert, and no telltale sound to warn those at ground level of the approaching danger. At ground zero, the V-2 typically spewed destruction across a wide swath, collapsing buildings and filling the streets with a cloud of churning debris. The V-2 left a crater about 30 feet in diameter, the signature footprint of Hitler's weapon of retaliation.[1]

Staver had come to Europe at the behest of Colonel Gervais Trichel, then chief of the rocket branch of U.S. Army Ordnance, a man who was keenly interested in investigating the enemy's new missile technology. With the anticipated collapse of Nazi Germany, Trichel had ordered the seizure of a hundred V-2s for evaluation, along with all technical documentation associated with the rocket. No less important to this top-secret mission was gaining access to Germany's talented cadre of rocket scientists and engineers. At the top of this list was the name of Wernher von Braun. Colonel Holgar Toftoy, an intelligence officer stationed in Paris, took charge of the trophy hunt, now given the name "Special Mission V-2." Clustered around Toftoy and Staver were a select group of intelligence officers, ordnance specialists, and civilian technical advisers. In the closing months of the war, these men soon discovered that they were not alone in the quest for the V-2 technology: Once the Soviets took Berlin, they, too, would display a keen interest in the German advances in rocketry.

The quest to gain access to the German rocket program was complicated by the war and its accompanying destruction. Major German cities, factories, and military bases had been routinely bombed. The ground assault on Germany proper in the spring of 1945 only augmented the ruin and disruption. The triumphant occupiers of Germany faced the daunting task of sifting through the debris of a defeated nation to discover what could be learned about the V-2 rocket and those who created it.

THE HIDDEN WORLD OF GERMAN ROCKETRY

The success of the V-2 resulted in no small measure from the leadership of rocket pioneer Wernher von Braun. Going back to the late 1920s, he had contributed both vision and organizational drive to the German rocket program. As a youth, von Braun had joined a band of rocket enthusiasts in Berlin. This small circle—with impressive engineering and technical skills—turned its attention to experimentation with liquid-fuel rockets.

During these formative years, the German rocket program was an exclusively civilian affair, animated by a vision of space travel and conducted on a shoestring budget. This civilian-controlled phase of rocket experimentation, however, proved to be short-lived. In 1932, the German army approached the fledgling rocket community with an offer of financial support for baseline research within the framework of military weapons development. Von Braun welcomed this overture from the army, arguing that such patronage was essential to move rocketry from the realm of a hobby to an arena for serious technological experimentation. The resulting alliance with the German military proved to be a fateful move, not just for von Braun and his fellow enthusiasts but for the history of modern rocketry.

Von Braun brought to the embryonic rocket program solid academic credentials, having earned a doctorate in physics at Berlin's Friedrich-Wilhelm University at the age of 23. His dissertation had focused on the theory of liquid-fueled rockets. While formally trained in physics, he had a flair for engineering, a rare combination of attributes that enabled von Braun to assume a role of leadership. He possessed a remarkable sense of timing, ultimately making peace with the Nazi regime, but also knowing when to act decisively to protect himself and his associates in moments of supreme danger. No one in the German rocket community, civilian or military, occupied a more elevated position of influence than von Braun.

Working with the German military (and later as part of Nazi Germany's war effort) inevitably raised questions about von Braun's motives and actions. The question of whether he was a Nazi sympathizer has prompted much debate. He did join the Nazi party in 1937 and, three years later, Heinrich Himmler's SS organization, the *Schutzstaffel* ("elite guard"). Both formal associations, his defenders have argued, should be viewed as expedient steps taken to sustain his research in rocketry. According to this interpretation, the formal links to the Nazi state did not reflect any personal ideology or, for that matter, any explicit sympathy for Hitler's political aims. In fact, von Braun expressed in small groups his disdain for the Nazi regime. In 1944, he was arrested and temporarily imprisoned by the SS on suspicion of not being wholly dedicated to the war effort; in the postwar years, this incident added some credibility to the argument that von Braun was a man at odds with the Nazi leadership.

However, the charge of complicity or passive acceptance of the Nazi regime would persist throughout von Braun's long career. In particular, Nazi Germany's genocidal employment practices in the war years intersected in controversial ways with von Braun's work. Slave labor was mobilized for the manufacture of the V-2 rockets in horrific conditions at Camp Dora, and more than 30,000 people died from forced labor, starvation, and illness. Both General Dornberger and von Braun, given their high rank, knew of the use of slave labor, even though they lacked any direct involvement in the administration of the camps. Von Braun and others, in varying degrees, were guilty at the very least of complicity. They did not challenge openly the use of forced labor, knowing it was vital to the German rocket program, and Camp Dora remains an inescapable black mark on the von Braun legacy.

Von Braun did not work in isolation, but at the epicenter of a remarkable community of rocket experimenters. Early on, he worked actively with the Society for Space Travel (*Verein für Raumschiffahrt,* or VfR), which served as an umbrella organization for these space visionaries. He also was an admirer of rocket pioneer Hermann Oberth, who had published in 1923 a pivotal theoretical work, "The Rocket into Planetary Space" (*Die Rakete zu den Planetenräumen*). Oberth's classic study offered a compelling basis to believe that space travel was technically feasible. Unlike literary visionaries Jules Verne and H. G. Wells, Oberth brought to his analysis a sophisticated understanding of aerodynamics and physics. Among his conclusions was the prophetic notion that a rocket with liquid propellants offered the optimal path for the future rocket development. Oberth, along with other prominent pioneers—Konstantin Tsiolkovsky in Russia and Robert Goddard in the United States—worked in isolation, at a time when there was no systematic body of theoretical knowledge on the principles of rocketry. From these lonely beginnings, Germany would become a fertile ground for many bold experiments with rocket-powered automobiles, sleds, and rail cars, which hinted at the future: the creation of a rocket to reach the upper atmosphere and beyond.

By the early 1930s, the VfR had established a test facility outside Berlin for a series of rocket experiments, attracting a talented group of engineers including Rudolf Nebel, Klaus Riedel, and the young von Braun. The rocket enthusiasts soon moved to a new locale for their experiments, a former munitions-storage complex in the Berlin suburb of Reinickendorf. Later dubbed the *Raketenflugplatz* ("rocket airport"), the abandoned site offered an ideal environment for experimental work: It contained concrete barracks, bunkers, and blockhouses surrounded by earthen walls 40 feet high and 60 feet thick. While the financial underpinnings for the Raketenflugplatz remained minimal, there was no lack of enthusiasm, as von Braun later observed.[2] Nebel, an engineer and former World War I pilot, took the lead in seeking out essential raw materials to sustain the rocket experiments. Soon the test site was filled with a lively group of engineers, electricians, draftsmen, sheet metal workers, and technicians— all drawn by the lure of rocketry.

Given the primitive nature of liquid-fueled rockets, many failures punctuated the work of the VfR, but the Raketenflugplatz nonetheless recorded some notable successes. The VfR enthusiasts took pride in the fact that they were able to launch arrow-shaped rockets that attained an altitude of 1,500 feet, then quite an achievement. These benchmarks soon attracted the attention of the German army, which viewed the rocket as a potential new avenue for weapons development. Germany had suffered a calamitous defeat in World War I (1914-1918). The resulting Versailles Treaty of 1919, which formally ended the war, placed burdensome restrictions on the German military. The victorious Allies used the treaty as a tool to severely curtail any future German program of rearmament and militarism. But there was a loophole to exploit: The treaty had not banned rockets or, for that matter, research on pilotless aircraft.

Seeing the loophole, Colonel Karl Becker, the army's chief of ballistics and ammunition, recruited members of the VfR to assist in the development of military rockets. Artillery captain (later general) Walter Dornberger, himself a rocket enthusiast, assumed oversight for the far-reaching experimental project. Von Braun and others in the VfR who joined the army program experienced a curtailment of freedom with the demise of the older civilian-sponsored rocket program. But that loss was offset by the realization that military sponsorship offered a more assured pathway to success. Now the army was in charge, subordinating the older vision of space travel to the real-world requirements of army weapons procurement. The previous goal remained a personal vision, but no longer the animating principle shaping German rocket research.

In 1932, the army shifted its rocket program to Kummersdorf, located 60 miles from Berlin, where a series of pivotal rocket experiments began. The army's goal was to build a long-range rocket, which would be the world's first ballistic missile. Such a missile, if operational, would be a formidable weapon, to be fitted with a one-ton warhead and capable of flying to a target between 100 to 200 miles away. The first major prototype, the A-1 (Aggregate 1) proved to be a dismal failure: It exploded on its initial launch at the Kummersdorf site. Never discouraged, the team then modified the essential design of the A-1 and fashioned a new rocket, the A-2. In December 1932, two A-2s, one called Max and the other Moritz (named after two celebrated German cartoon characters), completed impressive flights from an island on the North Sea, reaching an altitude of approximately 6,500 feet. The A-2 registered an important milestone, ensuring the continuation of army funding—at a higher level.

Once Adolf Hitler assumed power in 1933, he began to reorient Germany toward massive rearmament, in open defiance of the Versailles Treaty. The army rocket program benefited from the new government's largess toward weapons research. Though small and highly experimental, the rocket program soon received ample funding—all supplied in a context of tighter governmental controls and increasingly under a cloak of military secrecy. In early 1936, construction work began on the Peenemünde facility, situated on the island of Usedom on the Baltic Sea. This area, set apart for weapons research, housed the army rocket program, but also provided space for the *Luftwaffe,* the German Air Force, to begin a parallel program on a new pilotless aircraft, later to be known as the V-1.

Von Braun assumed the post of technical director of the army rocket-research center at Peenemünde. He and his team moved to the remote site in April 1937. Their first major project was the development of the new A-3 rocket prototype, which was 22 feet long and powered by an alcohol-oxygen engine with a rated thrust of 3,200 pounds.[3] Notwithstanding its advanced design, the A-3 proved to be a spectacular disappointment: In December 1937, three separate firings of the A-3 ended in failures. The rocket had a tendency either to explode suddenly or to fall into a spin and crash. The abortive A-3 launches placed new pressures on Dornberger to perfect the liquid rocket, and he realized that continued failures would lead to the cancellation of the army-funded program. The A-5, a companion to the A-3, finally proved to be successful, reaching an altitude of eight miles in 1938, at the time a stunning achievement.[4]

When World War II began in September 1939, von Braun's talented team settled on the A-4 (later V-2) as the most promising option to meet the army's requirement for an airborne weapon to hit distant enemy targets. The successful launch of the V-2 on October 3, 1942, set a milestone. Dornberger and von Braun briefed Adolf Hitler on the new rocket in July 1943, showing a film of the dramatic October launch. The V-2 offered a compelling prototype for a ballistic missile, given its size, range, and ability to carry an explosive warhead. However, the rocket remained a highly complex mechanism, costly to build, and never easy to operate. Albert Speer expressed some enthusiasm for the V-2, but he still retained some doubts about rocket's wartime utility. In the end, Hitler ordered the rocket into production.

DEPLOYMENT AS A VENGEANCE WEAPON

The initial V-2 strike on London took place on September 8, 1944, the first of 1,403 missiles to reach the city and other targets in southern England.[5] An estimated 5,400 people, mostly civilians, were killed in the V-2 bombardment campaign (with an estimated total of 12,685 killed by all V-1 and V-2 attacks). These casualty figures were modest when compared to the civilian deaths inflicted by the Allied strategic bombing campaign against major German cities. The V-2 had no counterpart in the Allied arsenal of weapons and was stunning in its impact. The famed American radio correspondent Edward R. Murrow, then posted in London, referred to the V-2 rockets as a manifestation of "malignant ingenuity" on the part of German science, more the harbinger of a future war than an effective weapon against the Allies in World War II.[6]

The Allies found it extremely difficult, if not impossible, to determine the place and time for a V-2 launch. Unlike the slow-moving V-1, the V-2 struck its targets from a very high altitude and at velocities in excess of the speed of sound. Even if spotted on the approach to a target, there was no effective air defense against the futuristic weapon. Nor could the launching pad be bombed. German firing crews fully exploited the mobility of the V-2 rocket. The rockets were transferred to their launch sites by truck.

The first step in the launch sequence was to place the V-2 on a *Meilerwagon,* an ingenious vehicle invented solely for the missile operations. Once attached to a special cradle, the rocket was raised by hydraulic means to a vertical position. A launching platform, a reusable 10-foot ring housed in a square frame, was then slipped under the rocket. The launch platform, supported by jacks on each corner, assumed the weight of the rocket, allowing the Meilerwagon to be removed. Each mobile unit required a command and control truck, assorted transport trucks, trailers, fuel storage tankers, and personnel carriers—typically 30 vehicles. Once a launch site had been selected, the German army cordoned off the area and removed all the local inhabitants; this procedure allowed for maximum security. Each crew took four to six hours to launch a V-2 rocket.

Before the launch the crew followed a precise routine to install guidance equipment, engine igniters, fuel, steering vanes, and other components. The V-2 required electrical power to operate, which was supplied by ground power and then batteries while the rocket was in flight. Given the dangers associated with any launch of a V-2, great care was taken to test fuel and ignition systems. The launch crew consisted of 20 soldiers, who were clad in special protective overalls and helmets for the fueling of the rocket. During the actual launch, the V-2 rose slowly off its metal platform, continued in a vertical ascent for about four seconds, and then inclined into its preprogrammed trajectory, controlled in flight by an inboard gyroscopic guidance system. The chosen angle of the trajectory, typically 45 degrees, determined the precise range for the missile. Engine shutdown came at around 70 seconds; at this point, the V-2 was moving across the sky at 50-55 miles altitude and at a velocity of 5,000 to 6,000 feet per second. The missile then made its descent, reaching the target five minutes after the launch. The impact on London and other cities was sudden and devastating. Even as the missile sped to its intended target, the V-2 firing crew quickly evacuated all their equipment and vehicles to avoid detection or counterattack by the Allies.

The BBC interviewed many Londoners who endured the first wave of V-2 attacks. Caught unawares at ground zero, these victims displayed shock and incredulity toward the radical aerial weapon. Visual sightings came after the V-2 struck its target, rarely before; most observers spoke of a "ball of light" accompanied by a "terrible crack." The V-2 appeared suddenly as if "out of the blue." There was no sense of the impending danger or possibility of defensive measures. "There was no alert," one Londoner told the BBC. "We had no warning at all. The first thing we knew was the sound of the explosion. One of the problems of this type of missile is that people do not and cannot take shelter. It might be a rocket or it might be a long-range gun. In any case, it must have traveled a terrific height into the atmosphere in order to come down as it did without us having any prior warning by sound of its approach."[7]

The deployment of Hitler's vengeance weapons was unnerving for the Allies, coming at a time when the Allied cause appeared to be on a fast track to victory. The operational use of jets, rockets, and flying bombs was a technological tour de force for Nazi Germany in its eleventh hour, but the new weapons did not reverse the tide of war. The number of V-2s striking Allied cities was relatively low, compared to the Allied strategic bombing campaign.

THE UNITED STATES WINS THE PRIZE

At the time of Staver's arrival in London in January 1945, the war in Europe had entered its final phase. The Allied armies in the west had liberated Paris and triumphed in the Battle of the Bulge. Now they were poised to continue the drive toward the Rhine and into Germany proper. In the east, the Soviets had invaded East Prussia and Czechoslovakia, setting the stage for the climactic and bloody battle for Berlin. The collapse of Nazi Germany was ratified in Berlin on May 8, 1945.

The final months of the war placed enormous burdens on Wernher von Braun and his technical staff at Peenemünde. The situation became extremely perilous with the approach of Soviet armies advancing from the east; the top-secret Peenemünde test facility stood directly in the path of the Soviet offensive. German refugees from East Prussia and other areas in the east clogged the highways. Defeat for Nazi Germany was now only a matter of time. Facing the inevitable, von Braun called a special meeting of his most trusted staff to discuss the urgent need to evacuate Peenemünde. This would include the wholesale removal of the technical staff, selected equipment, and key documentation to a safe place beyond the grasp of the invading Red Army. Von Braun took this occasion to voice what was apparent to all—that Germany, indeed, had lost the war. Their only option was to escape from Peenemünde immediately in order to preserve the essential core of their technical work. Saving the German rocket program, even in a rump form, served another goal, one that had animated their work from the beginning—the future possibility of space exploration. Among all the options, von Braun argued for surrender to the Americans, who alone had the means to preserve and ultimately advance their work. He also raised the prospect of working on explicit peaceful uses of rocketry in the coming postwar context.

As February approached, von Braun took the lead in planning for this extraordinary exodus from Peenemünde. The initial goal was to reach the Nordhausen area in the Harz Mountains of Thuringia, near the underground Mittelwerk V-2 factory. From this area in central Germany, the migration would continue 400 miles south to Bavaria. As a major in the SS, von Braun possessed critical links to the most powerful political and military entity within Germany. Yet this affiliation did not guarantee any measure of personal security. The SS might decide to execute the Peenemünde technical personnel as a way to deny the Allies access to the Nazi regime's most impressive secret weaponry. SS guards, in another possibility, might retain von Braun and his staff as a prize to use in any bargaining with the Allies. SS General Hans Kammler, an ardent and ruthless Nazi military operative, played an increasingly powerful role at Peenemünde in the final months. His erratic ways posed a potential threat to the escapees. But Kammler did not act to thwart von Braun or threaten the escapees; he would disappear under mysterious circumstances in the climactic days of the war. For von Braun, the desperate plan to move to Bavaria was worth all the risks. The group and their rocket legacy had to be relocated in a locale beyond the reach of the invading Soviets.

On February 17, after an interlude of desperate preparations, the escapees from Peenemünde departed for Bleichrode, a town near Nordhausen and their first stop on the southward journey. Their ultimate destination was Oberammergau in Bavaria. The group employed trains, trucks, and a barge; the latter was used to transport selected equipment and records. A total of 525 members of the technical staff with their families took part in the migration. Even as they departed, they could hear the roar of Russian artillery in the distance. While taking leave of Peenemünde, the evacuees used dynamite to destroy as much of the rocket installation as possible. These desperate attempts to deny the Soviets key technical equipment proved to be only a partial success. Once Peenemünde was occupied by the invading Red Army, there was a small residue of abandoned equipment to transport back to the Soviet Union.

To smooth the passage south, von Braun cleverly forged documents and bluffed his way to safety, passing through refugee columns and a chaotic gauntlet of checkpoints set up to seize deserters. Despite anxious moments, the passage to Thuringia was a dramatic success. Along the way, however, there was one near fatal incident: Von Braun endured severe injuries in a car crash.

While stopping at Bleichrode, the Peenemünde team took steps to hide their 14 tons of records in an abandoned mine—the invaluable paper trail of more than a decade of research and experimentation. The long journey ended on April 4, when von Braun and his colleagues reached a remote ski lodge in Bavaria. It would be here that von Braun, Dornberger, and the technical elite from Germany's rocket program eventually surrendered to the American army.[8]

At the war's end, the Nordhausen area—the locale for the Mittelwerk factory and the hidden documents—became the main objective for an intense competition among the Allies to seize the surviving V-2 technology. The Nordhausen region in April-May 1945 mirrored all the chaos that gripped Germany in its hour of defeat. Charles Lindbergh, the famed American aviator, has left one of the most vivid accounts of the time. Lindbergh came to occupied Germany as a member of the Navy Technical Mission, which allowed him access to some of the key Nazi installations associated with advanced weaponry. He arrived in Nordhausen shortly after advance elements of the American army had secured the Mittelwerk underground factory. Traveling in a convoy of jeeps and trucks, Lindbergh's group reached areas still tenuously under the control of remnants of the German army—in his words, "an army surrendered, but with its guns and discipline still intact." At one crossroad, he remembered, his team confronted an entire regiment on the march—"trucks, cannon, and thousands of rifled infantrymen strung out at right angles to our line of travel. A seemingly endless column of green-clad German soldiers moving in one direction; a single American jeep rolling up against it, like a fly buzzing at the body of a serpent."[9] This was the same chaotic scene von Braun had encountered just days before Lindbergh's arrival.

Later, Lindbergh made a dramatic visit to the Mittelwerk factory, where he inspected an assembly line of V-2 rockets, abandoned by the Germans. He drove a jeep right into an underground factory that had been blasted out of solid rock. He observed the facility up close: "Hundreds of subassemblies of V-2s were lying on flatcars or were scattered over the ground: nose cones, cylindrical bodies, and big fuel tanks for liquid oxygen or alcohol. A number of tail sections, shining and finned, were standing on end like a village of Indian teepees. . . . We left our jeep and walked into a side tunnel, which contained a production line for V-2 engines. In another tunnel we saw an entire rocket assembled. . . . We walked back and forth through the cross tunnels. There were miles of them, lighted, deserted, silent."[10] Even as he observed the rocket debris field at Mittelwerk, Lindbergh had no idea that the American army would soon engage in a high-stakes gamble to retrieve this technological legacy of Nazi Germany—at the expense of the Soviet Union.

As head of "Special Mission V-2," Colonel Toftoy aimed to secure enough components to assemble 100 V-2 rockets. He deployed Major William Bromley, a trusted high-energy assistant, to Nordhausen to procure these components. The remains of the Nazi rocket program were to be shipped to the United States for reassembly and testing. No less important, the special mission team wanted to confiscate all relevant documents and to interview technical experts associated with Peenemünde. Toftoy and his team, however, faced a major complication associated with the execution of their search. The Nordhausen area, as all of Thuringia, had been designated to be placed under Soviet control, effective June 1, as part of the Yalta agreement dividing occupied Germany into four zones. They had to race against time to find and evacuate the V-2 hardware and records before the arrival of the Soviet army.

Bromley faced the daunting challenge of securing the V-2 components in a relatively short time frame, since formal approval did not arrive until May 22. Never at a loss for improvisation, Bromley hired former slave laborers to assist in the top-secret operation. As the hired workers cleared the vast tunnel complex of V-2 parts, Bromley managed to secure the requisite transport—trucks and railcars—to evacuate the treasures of Mittelwerk. To achieve this end, he had to requisition trucks deep in the rear, as far away as Cherbourg, France. A steady stream of shipments eventually made their way to Antwerp, Belgium, where they were crated and loaded on awaiting Liberty ships for transport to the United States. The precious cargo from Nordhausen was destined for White Sands, New Mexico, where the army planned to reassemble and test the V-2 rockets.[11]

Major Staver, also part of the Toftoy team, faced a more difficult challenge in seeking out the buried documentation for the V-2 rocket. Staver's task was to find the abandoned mine where the documents had been hidden by the von Braun team. This was no easy task, given the chaos of the war and the general hostility of the local population. Following a tip, he was able to locate the buried documents in a mine near the small village of Dorten, 30 miles outside Nordhausen. On May 26, on the eve of the Soviet arrival in Nordhausen, Staver managed to requisition six large trucks to remove the documents to the American zone.

THE SOVIETS MAKE THEIR MOVE

The Soviet Union had been aware of the V-2 missile technology during the war. At the closing of the war, however, the Soviets possessed only an incomplete picture of this advanced rocket technology; obtaining it was a priority. However, rocketry was just one part of a concerted campaign to confiscate a vast array of industrial and scientific treasures in occupied Germany. Once the Soviets began an intense campaign to seek out remnants of the V-2 technology, they were frustrated to learn that the Americans had already removed much of the most important finds—both hardware and human expertise.

Boris Chertok, a specialist in rocket control, guidance, and communications systems who worked at NII-1 (Scientific Research Institute— No. 1) of the People's Commissariat of the Aircraft Industry, has provided one of the best firsthand accounts of how the Soviets sought out the vestiges of the German rocket program. In his lively memoir, titled *Rakety i lyudi* ("rockets and people"), Chertok reveals the slow and often chaotic Soviet response to the trophy hunt for the V-2 rocket.[12] The occupation of Germany created great confusion on the Soviet side, because a variety of Soviet industries, along with a range of military departments, participated in the quest for useful trophies to take home. Chertok had been associated with the Soviet aviation industry as an engineer, and initially his bosses were more interested in Nazi Germany's advanced jet engine technology, which offered an obvious avenue for modernization. By contrast, the rocketry of Nazi Germany—if stunning and unique—appeared more futuristic, even impractical to some military planners. And, no less important, there was a dilemma over who should assume control of the secret V-2 spoils. In the absence of an existing missile program, it became quickly apparent that a new governmental entity would be required to oversee the development of Germany's advanced rocket technology.

Chertok had first learned of the V-2 missile in 1944, when the Soviets invaded Poland and recovered remnants of the new rocket from an abandoned German rocket test site. Collaboration with British intelligence quickly revealed the capabilities of the secret weapon. Chertok's own NII-1 research institute had collected data from Poland. By 1945, Chertok and others were arguing that a systematic search be conducted for the V-2 rocket, in the hope of gaining first access to this futuristic technology. Those in charge of the aviation industry, however, did not share Chertok's sense of urgency; for them, the main task at hand was the collection of advanced German machine tools and jet engines. As an engineer, Chertok took a professional interest in German advances in guidance technology, rocket engines, and radio-control techniques. He, too, marveled at the ability of the Germans to create a powerful liquid-propellant rocket engine, one that could perform at supersonic speeds and be adapted as a missile to bomb Allied targets.[13] As the war drew to a close, Chertok realized how critical it would be for the Soviet Union, in his words, "to get to the front and be the first to seize the intellectual war spoils of rocket technology."[14]

Chertok flew to war-ravaged Germany in late April 1945, as part of a group led by General Nikolai Petrov, one of several groups seeking Nazi Germany's technological trophies. After the war, on June 1, 1945, Chertok made his first visit to Peenemünde. He recorded in his memoir that he had been drawn to the test site by "engineering curiosity" and "a sense of duty to our country." Flying over the area, he was struck by the sheer beauty of the Baltic coast and the outward appearance of Peenemünde as a resort. On closer examination, he observed the desolation of the former German rocket facility, the destroyed buildings, laboratories, test stands, and factory complexes. The Germans had systematically dismantled the facility in advance of the Soviet occupation, even down to the removal of machine tools and equipment. Still intact, though, were the bunkers, the excellent roads, and the elaborate network of power and signal cables—appearing now as the skeletal remains of Peenemünde. Prior to their evacuation, the Germans had done an effective job of denying the Soviet conquerors the treasures of Peenemünde.[15]

Later, on July 14, Chertok arrived with his team in Nordhausen. On this trip Chertok inspected the center of V-2 production at the Mittelwerk facility. He noted that the city, in the immediate aftermath of the war, revealed the devastation of Allied bombing. He visited Camp Dora, which had been the home for the thousands of slave laborers who worked on the Mittelwerk assembly lines. At this juncture, Chertok noted, the notorious death camp had been cleared, with the dead buried and the survivors offered food and shelter. One former prisoner at the camp delighted Chertok with the offer of a gyroscope from a V-2 rocket, an artifact he had hidden and concealed over many months. Once Chertok reached the underground Mittelwerk, he saw that the Americans had been efficient in removing the most valuable partially assembled V-2s and a huge amount of rocket components. A local German engineer, who had worked at the factory, informed Chertok that the assembly lines had functioned up to the last days of the war, producing 35 missiles a day. He said the Americans had taken away the bulk of the rockets, but he still believed the Russians could sift through the debris and reconstruct as many as 20 missiles. This was the Soviets' consolation prize.[16]

In time, Chertok took up residence with his team in the Villa Frank, a former mansion of a wealthy family in nearby Bleichrode, where von Braun had temporarily lived in the final days of the war. Chertok was overwhelmed with the luxury of this house: its large rooms, marble staircase, gilt-framed pictures, and sumptuous appointments. Chertok and his small group worked diligently to reconstruct the flight-control system for the V-2 rocket. To achieve this end, they organized the *Institut Rabe* (an acronym for *Raketenbau und Entwicklung Bleichrode*, "missile construction and development in Bleichrode").[17] The organization aimed to coordinate the research and mobilize German technicians to provide needed expertise. Chertok even made an abortive effort to lure Wernher von Braun to the Soviet side. One successful step, however, was the establishment of a research outpost at Lehesten, where a test stand for the V-2 rocket engine had survived intact. Here the Soviets set up a program of research on propulsion technology. Valentin Glushko, destined to become a major figure in the Soviet space program, led the team at Lehesten, a project that would remain in Germany until January 1947.

How the Soviets organized their rocket program was a complicated affair, especially in the immediate aftermath of World War II. Chertok's Institut Rabe was co-opted in August 1945 by the chief artillery directorate. Later, General Lev M. Gaidukov, who had overseen the development of the Soviet Katushka battlefield artillery rockets, organized an interagency commission to deal with the V-2 technology. There was no small amount of bickering and confusion in this process. One key goal for the Russians was to seek out ways to put the V-2 rocket back into production and to recruit former German technicians to assist in the studying of advanced weaponry. The Soviet campaign to exploit fully the vestiges of advanced German weaponry, particularly the V-2, was not always a well-ordered endeavor. More than one organization took shape to coordinate the process of collection, research, and testing. Given the resulting overlap, Stalin created a "Special Committee No. 2" in May 1946 to oversee a coherent program of research on rocketry.[18] This organization would organize the production of Soviet missiles in the post-war years, making the V-2 the baseline technology. As a consequence, Stalin ordered the rocket program distributed among various industrial sectors with a special governmental committee to oversee the research and development work. From the midst of this organizational change, Dmitriy Ustinov, the Minister of Armaments, soon emerged as the leader of the Soviet Union's ballistic missile and space programs.[19]

The Soviets' key weakness was their inability to attract high-ranking and talented German scientists associated with the rocket program at Peenemünde. The major exception was the recruitment of Helmut Gröttrup, a former assistant to von Braun and a leading figure at Peenemünde, where he was the director of the laboratory for guidance control and telemetry. Rejecting the option of joining von Braun in Bavaria or, more important, surrendering to the Americans, Gröttrup—with his family—decided to join the Soviets at Institut Rabe in September 1945. He accepted an affiliation with the Soviets in no small measure for the material inducements offered by the Russians and the prospect that he could remain in Germany. Other key figures were also persuaded to join the Soviet side: aerodynamicist Werner Albring, design engineer Josef Blass, guidance and control expert Johannes Hoch, gyroscope specialist Kurt Magnus, and propellants chemist Franz Mathes, among others.[20]

The dependence on German specialists reflected the backwardness of the Soviet rocket program. The Germans who collaborated with their Soviet overlords enjoyed many privileges and financial rewards, but their status remained a highly subordinate one, always governed by the Soviet mania for secrecy and the ubiquitous presence of the secret police. They were strictly relegated to a consultative role with minimal independence. Working in the Russian rocket program was fraught with peril, as shown in October 1946, when Stalin's secret police forcibly deported selected German technical specialists, including Gröttrup, to the Soviet Union. An estimated 2,200 German experts in aviation, nuclear energy, rocketry, electronics, and radar technology were compelled to work in a variety of Soviet industrial entities. This traumatic deportation contrasted sharply with the fate of Wernher von Braun and the Peenemünde specialists who opted to work in the United States.

Irmgard Gröttrup, the outspoken and assertive wife of Helmut Gröttrup, noted with bitterness in her memoirs that the Soviets had promised the German experts they would never be sent to Russia. Instead, she found herself and her family on a train to the east: "Their grin was as friendly as ever. Indeed they even made a few promises: a flat much larger and much nicer than ours, a life without any restrictions, a life in a magnificent city amongst grand people. The only thing they couldn't promise was when we should see our country again. . . . At one point, simply to be free for a moment, I tried to get out through the back door. Impossible! The barrel of a gun—a broad face: 'Nyet.'"[21] The Germans with specializations in rocketry were assigned to a variety of research and test facilities in and around Moscow. They labored in difficult conditions for several years, and some were finally allowed to return to Germany at the end of 1950. Gröttrup and six others were the last of the German scientists to leave in the mid-1950s. Their release signaled that the Soviet rocket program no longer required the tutelage of these former Peenemünde veterans.

In September 1945, Sergei Pavlovich Korolev, the future leader of the Soviet space program, made his first appearance in occupied Germany. Korolev was a dynamic figure, with an impressive portfolio in both aviation and rocketry. An engineer by profession, Korolev took an early interest in rocketry in the 1930s. He was arbitrarily arrested in 1938 during the purges. Accused of subversion and treason, Korolev was tried, convicted, and sentenced to 10 years of hard labor in Siberia. At the infamous Kolyma gold mines in eastern Siberia, Korolev labored under the most severe conditions, working in extreme cold, often without proper clothing. His health declined dramatically; he lost all his teeth and then developed a heart condition. In September 1940, aviation designer Andrei Tupolev arranged for Korolev to be transferred to his "sharashka," or police-run workshop, in Moscow. This timely intervention saved Korolev's life. He went on to survive the war years, though plagued by a chronic heart condition. Soon he emerged as a valued engineer in the Soviet drive to exploit the V-2 rocket technology.[22]

Sent by the People's Commissariat of Armaments in Moscow, Korolev arrived in Berlin to study German rocketry. As with many civilian experts, he was given the rank of lieutenant colonel in the Soviet army to legitimize his activities in the former war zone. While in Germany, Korolev worked diligently to familiarize himself with the V-2 design. In October, he viewed an actual V-2 launch by the British at Cuxhaven on the North Sea. At the Institut Rabe, Chertok was struck with Korolev's intelligence and personal drive. Not only did Korolev prove to be a thoroughgoing student of the advanced V-2 technology, but he also demonstrated an impressive skill at refining the basic rocket design. He created a special task force named "Vystrel"; the portfolio for the group included research on pre-launch preparation equipment and flight-mission control.[23] The sojourn in Germany heralded a new phase in Korolev's life. He thrived with his new freedom. By 1946, Korolev assumed a new job, as chief designer for long-range missiles, with the NII-88, a research institute in Moscow responsible for the production of Soviet missiles based on the V-2 technological template.[24]

By the summer of 1947, the Soviets had succeeded in reassembling a small number of V-2 rockets for tests at Kapustin Yar, located 56 miles southeast of Stalingrad (present-day Volgograd). Here Korolev played a key role, supported by Gröttrup and the other interned German technicians. The test range was remote, situated in a desert, but adjacent to a rail link to Stalingrad. Conditions at Kapustin Yar were austere, with no permanent structures, few amenities, and only primitive housing for the staff—with many forced to live in tents or railcars. The weather was extreme and unforgiving, and for the unwary, poisonous snakes and tarantulas added a note of peril to life there.[25] The first V-2 launch took place on October 18, as a partial success, but the rocket disintegrated upon reentry into the atmosphere. A total of 11 rockets were launched, with five classified as successes, but the others deviated from their targets, exploded, or experienced some sort of technical failure. This work at Kapustin Yar was important for the future of the Soviet space program. By contrast to the

Americans, the Russians possessed only a handful of workable V-2s, and they made excellent use of them. The launches at Kapustin Yar allowed the Russians to engage in upper-atmosphere research for the first time. The program ended with the decision to manufacture the R-1 rocket. It would become the Soviet Union's first rocket design in the postwar years.

THE AMERICANS TEST THE V-2 AT WHITE SANDS

The men who had designed the V-2 rocket constituted an important war prize. The United States War Department through its Joint Intelligence Objectives Agency (JIOA) set in motion a campaign to recruit the most talented specialists from Peenemünde. Such specialists, it was thought, were necessary to assist the United States in the testing and development of the V-2 rocket. The secret program was first called "Operation Overcast," but it eventually acquired the name "Operation Paperclip." This peculiar code name, as the story goes, derived from the practice of placing paperclips on the immigration forms for key scientific recruits from Germany. The twofold purpose of Paperclip was to recruit the best German specialists on rocketry and simultaneously deny this same technical expertise to the Soviets, who quickly emerged as a major rival to the United States for war spoils. Nearly 500 German recruits with their families participated in Operation Paperclip in 1945, a group led by von Braun. The Germans were sent to the U.S. Army's White Sands Proving Ground, near Fort Bliss, in New Mexico. Here they were reunited with the 100 V-2s seized at the Mittelwerk factory in May 1945.

The entire program was controversial from its inception, prompting criticism in the War Department among many who opposed the importation of any individuals associated with the Nazi regime. The recruitment and processing of these German rocket experts took place outside routine State Department review or approval. The first contingent of Germans reached the United States in November 1945. Personnel files of many of the Germans approved for Paperclip did reveal a wide variety of associations with the Nazi-controlled government. President Harry S. Truman had approved the Paperclip project in August 1945 with the understanding that no one with a record of political activism would be admitted to the United States. As it turned out, this high standard was not consistently maintained, in part because the German rocket experts were so important for American national security. Much of Operation Paperclip would remain classified in the decades after World War II.

Von Braun emerged as the dynamic leader of the transplanted community of German rocket scientists. His leadership skills had been established in Germany, notably in the dangerous escape from Peenemünde and subsequent surrender to the Americans. He came by this trait naturally, since he was born into a Prussian Junker family. He made a dramatic impact on all those who encountered him: He was a tall man (5 feet, 11 inches) with thick blond hair, a square jaw, and striking good looks. He was known for his athleticism, booming laugh, and skills as a conversationalist. Friends and strangers alike were taken with his personal charm, his fluency in several languages, and his diverse cultural interests. At Peenemünde, for example, he participated in a chamber music group, and took great delight in playing both the piano and cello. Throughout his life, he proved to be an avid reader with myriad interests in philosophy, religion, geography, and politics. If frank and direct in his dealings with those around him, he also was devious, ruthless, and capable of astute political maneuvering in the high-risk world of Nazi Germany. For all of his pragmatic skills, however, he had never stopped embracing the dream of space exploration, a fixation that added a prophetic dimension to his career.[26]

Life at White Sands for von Braun and his colleagues was difficult and challenging, filled with equal measures of professional attainment and bouts of personal unhappiness with their uncertain status. On the one hand, they launched nearly 70 V-2 rockets, far exceeding the parallel program of the Soviets at Kapustin Yar. The White Sands tests were highly successful, offering a floodtide of scientific and technical data.[27] On the other hand, though, the Germans lived in primitive barracks, in a complex with rudimentary support facilities: a mess hall, administrative and supply buildings, and a recreation club. Initially, the men working at White Sands did not have passports or documents to allow them to move freely beyond the army base. At least there were no fences, since the desert itself offered a barrier to the outside world. Beyond the proving grounds were the San Andreas and Sacramento mountains. The summer days were hot and stultifying. The nearest urban center was El Paso, Texas, where the Germans were allowed to visit on occasion.[28]

Daniel Lang, a writer for the *New Yorker*, visited White Sands in 1948 and has left a vivid image of the V-2 test program: "White Sanders whose outlook is anything but downtrodden—[are] highly enthusiastic men who are glad to be almost literally shooting for the moon." The rocket firings, Lang reported, took place every few weeks, and often these launches attracted high-ranking military figures. Those civilians who lived near White Sands had grown accustomed to "whizzing flashes over the desert waste." The proving ground is 40 by 90 miles, a reservation of "endless sand" filled with rattlesnakes and other natural hazards. Those working at White Sands, Lang reported, were a mix of German veterans from Peenemünde and assorted American experts drawn from the military, industry, and academia. On launch days a "gala" atmosphere reigned, with civilians clogging Route 70, which ran through the military reservation. The public spectacle of the V-2 firings was mirrored in the crowds converging on White Sands: "Buses loaded with Boy Scouts, R.O.T.C. cadets, college students, National Guardsmen, and delegations from chambers of commerce and civic clubs. . . ."[29] White Sands, with its relative openness, contrasted sharply with the hidden nature of the Soviet V-2 rocket experimentation.

The testing of the V-2 at White Sands possessed a clear scientific purpose. Rockets launched at the proving grounds were routinely fitted with instrumentation to study solar spectroscopy, cosmic rays, and the measurement of pressures and temperatures in the upper atmosphere. There were problems associated with this pioneering research, in particular because the trajectory of the V-2 allowed for only short interludes at extreme altitudes. Engineers gave careful attention to utilizing the V-2's warhead compartment to house instruments. Special access panels were created to install and adjust the variety of scientific instruments used in the program. Cameras were mounted on several V-2s launched at White Sands, taking photos of Earth from altitudes of up to 100 miles.

"The rocket," observed Ernst H. Krause in 1947, "has opened the door to vast regions of space which at present are known to us primarily through the astronomer's telescope."[30]

The momentum of the White Sands program continued well into the 1950s. A Viking rocket launched in May 1954 returned pictures from 158 miles above Earth, showing El Paso, the winding Rio Grande, several railroad lines, clouds, and, of course, the surrounding desert. Some of the pictures, taken at an oblique angle, showed the curve of the Earth and even the blackness of space beyond the planet's atmosphere.[31]

The images, which were released to the public, were precursors of several vital aspects of the coming exploration of space. First, they showed the potential of space photography to peer into what intelligence agencies sometimes euphemistically referred to as "denied territory," the Soviet Union. Second, the photographs demonstrated that views from space could enable meteorologists to see, for example, hurricanes and storms forming at sea well before they reached land—in time to warn those in the path of adverse weather. Finally, and perhaps most important in the long run, these early space photos were steps in a journey that eventually would enable humans to view the entire planet and much else beyond and to understand Earth's place in the solar system.

Nor did such early "peaceful" space experimentation stop at sending cameras into space. Monkeys, dogs, and mice were launched on short, suborbital rides. Beginning in mid-1948, several V-2s carried rhesus monkeys and mice on flights in the United States, with mixed results. Some of the animals survived, while others did fine on the ride but were killed on impact at the end of their flights. Such tests continued into the early 1950s.[32]

As for rocketry development itself, the original testing of the V-2 was part of the Hermes ballistic missile program. Efforts were made to improve the basic design of the V-2 and experiment with two-stage rocket configurations. On February 24, 1949, one V-2 was modified to fit a second stage called the Bumper WAC rocket. In this pivotal experiment the redesigned rocket reached an altitude of 250 miles.[33] This milestone signaled the advent of multistage ballistic missiles, which would be developed over the next decade.

In 1950, von Braun and his team of transplanted German rocket scientists left White Sands for the Redstone Arsenal in Huntsville, Alabama. The German specialists greeted this decision with enthusiasm. The chance to escape the isolation and austerities of White Sands raised morale. The several years of experimentation with captured German V-2 rockets had run its course. It was time to move on. The Army facility at Huntsville housed a pair of former arsenals, including one used in World War II for the manufacture of poisonous gases and other chemical munitions for artillery projectiles. By 1950, these two separate arsenals had been combined into a single facility and selected for the newly created Army Ordnance Rocket Center. A friend recalled von Braun's excitement when he returned to White Sands after his first visit to Huntsville before the move: "Oh, it looks like home! So green, everything is so green, with mountains all around."[34] Over a period of several months, beginning in April 1950, 115 of von Braun's former Peenemünde team members and their families moved from the desert of New Mexico to the rolling hills of northern Alabama. They were joined by others from the White Sands facility, including a few hundred General Electric Company contractor employees and a group of U.S. Army draftees holding math, science, and engineering degrees.[35]

Von Braun was 38 years old when the move took place; he arrived with his wife, Maria, whom he had married in 1947, and their young daughter, Iris Careen. Von Braun and his colleagues were invigorated by the relocation itself as well as by a renewed sense of mission. Their new home was a quiet rural town of 16,000 in the Appalachian foothills of the Smoky Mountains. Huntsville called itself "The Watercress Capital of the World"—later it would be known as "The Rocket City"—and the new residents received a friendly welcome, the only exceptions being a few locals who had lost relatives or friends in the war.[36]

In the 1950s von Braun worked on the Redstone rocket, becoming a creative contributor to the Army's rocket program. While working at Huntsville, he would also emerge as the pivotal figure who gave voice to the American space program.

WHEN GOOD ROCKETS
GO BAD

The more complex a device, the more ways it can fail. Rockets concentrate high complexity with extreme forces, making them very dangerous when they fail—whether it is a burst seam, a shorted control circuit, or a foreign particle trapped in a valve, most "failure modes" lead to catastrophic results.

Rockets do sometimes "fizzle out" and simply fail to launch, as in the first launch attempt from Cape Canaveral in 1950, when "Bumper 7" simply sat immobile on the pad at the launch command due to a valve corroded by the salty coastal air. But with all their complexity, the tremendous pressures and temperatures involved, and the energy of their propellants, a great many rocket failure modes result in devastating explosions, which not only destroy the vehicle, but can reduce a launch complex to scrap as well.

Early launch pads at Cape Canaveral were built in pairs since planners took it for granted that an explosion would eventually wreck at least one of the pads. The Thor missiles at Launch Complex (LC) 18 lived up to these expectations and then some, blowing up at both the A and B pads in a series of launch failures. The Navaho ramjet cruise missile detonations blew parts and equipment all over LC-9. Pad 18A was refurbished for the first Vanguard launch, carried on live television, but the rocket rose for only two seconds before falling back on the stand and bursting into a fireball. The Atlas missile series, intended for eventual use by the astronauts, racked up an alarming explosion rate on the pad and in the sky until design issues were solved.

Launch control centers were built like armored bunkers to protect them in case of launch disasters. Igloo-like concrete shells covered many of the control centers along "Missile Row" at the Cape. Mere concrete and armor, however, would not be enough in the face of a colossal moon rocket explosion. The launch control center at LC-39 had to be located more than *three miles* from the pads to survive the blast radius of a Saturn V explosion. Fortunately, not one of von Braun's Saturn rockets ever failed.

Von Braun's close-knit rocket team developed strict and cautious countdown and launch site protocols that recognized the tremendous danger inherent in rockets. Despite dozens of disasters, no Cape personnel were injured or killed by rocket explosions.

The Soviets were less fortunate. Late in 1960, a new intercontinental ballistic missile sat on its pad at Baikonur Cosmodrome in Kazakhstan. The R-16 ICBM used propellant chemicals that were so potent and reactive that they needed no spark to ignite: "Hypergolic" in nature, they burned fiercely and instantly on contact with each other. Highly toxic and highly corrosive, the fuel mixture was called "devil's venom" by the Soviet technicians. On October 24, 1960, the R-16 stood on the Site 41 launch pad fully fueled, while a large group of harried technicians worked in the gantries under the eyes of their frustrated engineer-supervisors and the personal gaze of the commander of the Soviet nuclear missile program, Marshal Mitrofan Nedelin. The standard safety protocol required draining the R-16 of fuel before allowing personnel near it to continue work, but the de-fueling would seriously delay the intended launch schedule, and so Marshal Nedelin had ordered the launch crew out to the pad.

An electrical error opened the valves of the rocket's second stage engine. The hypergolic chemicals ignited almost instantly, firing into the first stage below it and detonating that fuel tank as well. The men on the gantries were incinerated. Flames poured out of the rocket as the ground crew ran for their lives, showered by the "devil's venom," burning and choking to death on the toxic gas as they struggled to tear off their burning clothes. A wire fence entangled some of the men as the fireball expanded to engulf them. Others found too late that an asphalt pathway had melted into tarry glue, which held them like flypaper. Temperatures may have reached some 3,000°F. The whole scene was recorded by automatic cameras set to film the glorious launch.

We will never know all the details owing to Soviet secrecy, but the death toll was about 130 people, including the marshal at his ringside seat. This horrific accident killed not only highly trained pad technicians but also many of the Soviet Union's best rocket engineers, a substantial loss of expertise at a critical point in the space race. The Nedelin disaster illustrates the danger of overriding engineering judgment in this inherently dangerous field. The Soviet system allowed for autocratic military power to run roughshod, and the launch team at Site 41 paid the price. Only hard-earned experience, meticulous engineering, and tremendous care makes rockets anything but very expensive bombs. ■

SPACE AND THE COLD WAR

The decade of the 1950s signaled a period of deepening tensions between the United States and the Soviet Union. Americans felt a growing alarm over the perceived global threat of communism. The fear of domestic spying became a powerful force in American life in the postwar era, as shown in the celebrated case of Julius and Ethel Rosenberg, who were executed as Soviet spies. In Eastern Europe, the Soviet Union consolidated its power with the creation of communist-run states, prompting Winston Churchill to refer to an "iron curtain" descending across the continent. In 1948, a coup brought Czechoslovakia into the Soviet bloc. The Berlin Airlift that same year successfully blunted a Soviet move to isolate and seal off the former German capital city, then under four-power occupation. During these turbulent times, the administration of President Harry S. Truman used the Marshall Plan to assist European democracies in economic recovery, offering financial aid with the aim of assuring political stability. No less important, the so-called "Truman Doctrine" mobilized direct

OPPOSITE: *A U.S. Army Redstone missile, developed by von Braun and his team, is launched at Cape Canaveral.*

economic and military assistance to assist Greece in fighting a communist insurgency. These successes, however, were dwarfed in 1949 with the triumph of communism in China under Mao Zedong. The decision of North Korea to invade South Korea in 1950, which sparked the Korean War, only deepened fears of continued communist expansion.[1]

The Soviet Union appeared to most Americans as a new and formidable military threat. By this time the Communist state possessed the atomic bomb (first tested in 1949), a developing fleet of long-range bombers, and an active program of missile development, the latter based on their own exploitation of the German V-2 rocket technology. When Dwight D. Eisenhower assumed the presidency in January 1953, he made a renewed effort to counter this extraordinary threat from the Soviet Union and its allies. Eisenhower sought international agreements to ban the testing of nuclear weapons, pursued new initiatives to gain intelligence on Soviet technology and military intentions, and began to expand experimental work in missile development. Eisenhower faced the dilemma of providing for a robust national defense and at the same time maintaining the delicate peace with the Soviet Union.

SPACE IN THE POPULAR IMAGINATION

Despite the pall cast by Soviet military might, the decade of the 1950s would also give birth to a supernova of imaginative thinking about space travel, in books, magazines, radio and television, and the cinema. In contrast to the 19th-century visions of Jules Verne in *From Earth to the Moon* (1865) and H. G. Wells in *War of the Worlds* (1898)—which were shaped by sheer fantasy—this new generation of space enthusiasts envisioned the dream of space travel from the perspective of modern science and technology. These latter-day visionaries—a diverse group of artists, scientists, philosophers, rocket engineers, and science-fiction writers—offered plausible ways for humans to escape Earth's gravity and explore outer space. Civilization, in their view, was on the cusp of a new age of exploration. And the script for this epoch was clear in broad outline: building an orbiting space station, landing on the moon, and trekking to Mars and beyond.

The prospect of future space travel had been heralded by the V-2 rocket. Modern rocketry soon held promise for both military and civilian uses. The idea of reaching outer space appeared concrete and imminent, no longer a fantastic dream or the stuff of science fiction. The space age had arrived.

The first important salvo of the new consciousness came in 1949 with the appearance of a small book, *The Conquest of Space*, written by Willy Ley with illustrations by Chesley Bonestell.[2] Robert A. Heinlein, the celebrated science fiction writer, reviewed the illustrated book in *The Saturday Review,* where he greeted it as the "Baedeker of the Solar System." For Heinlein, the book represented the "next best thing to interplanetary flight." Ley, an émigré German engineer and popularizer of space travel, had come to the United States in the 1930s, after participating briefly in Germany's pioneering rocket community. Along with Bonestell's imaginative paintings, Ley wrote a vivid text showcasing futuristic scenes of the human exploration of the cosmos. Putting this remarkable new book in proper perspective, Heinlein noted that atomic energy had made a recent and profound impact on his generation, but *The Conquest of Space* suggested that yet "another revolutionary thing [was] about to happen to the human spirit: space travel."[3]

In March 1952, *Collier's,* a national weekly boasting a circulation of four million, inaugurated a series of six articles on rocketry and space exploration. George Manning, the magazine's managing editor, played a key role in conceptualizing the series, titled "Man Will Conquer Space Soon!" Later, Cornelius Ryan joined the project as the chief editor. The origins of the *Collier's* series had in part grown out of a symposium on space travel at New York's Hayden Planetarium, launched the previous year by Ley, fittingly on Columbus Day. For the magazine, the undertaking was a bold move, representing a major breakthrough in the popularization of modern rocketry. A number of eminent scientists and rocket engineers participated as consultants and writers, most notably Wernher von Braun, then working in the Army missile program at Huntsville, Alabama. Others joined von Braun: Willy Ley, Harvard astronomer Fred Whipple, physicist Joseph Kaplan, and Heinz Habler, an expert on flight medicine, among others. The assemblage of such technical expertise gave the series an air of authority and relevance.[4]

Collier's, with its vast popular readership, took pains to give the sequence of articles on space a stunning visual appeal. Here imagination blended harmoniously with existing science and rocket technology to fashion a compelling view of the future. Many who read the *Collier's* articles were awestruck by the illustrations: the celestial vistas, the elaborate architecture (often with cutaways) of orbiting space stations, the precise and technically informed renditions of future multistage rockets, and the imaginative, often cinematic, portrayal of future missions to the moon and Mars.

One image in particular captured the attention of readers: the vision of a future space station, featuring a ferry rocket and astronauts walking in space. The huge three-deck station, as described in the magazine, was the shape of an enormous wheel, measuring 250 feet in diameter, and orbited at an altitude of 1,075 miles. It rotated at three revolutions per minute, to simulate one-third of Earth's gravity. The station would be a permanent fixture, with its elaborate living and work areas, self-sustaining solar-powered boiler and turbine plant, sophisticated navigational and surveillance instruments, and docking bay for visiting spaceships. Most important, the futuristic space station would become a launching pad for missions to the moon and Mars.[5]

On October 18, 1952, *Collier's* published the second installment in the space program series, in this case an anthology of three articles under the general heading "Man on the Moon." In the accompanying editorial, "Next Comes the Moon," the magazine took note of the fact that the initial installment in the series, published the previous March, had coincided with an important international conference on space exploration in Stuttgart, Germany. More than 200 scientists had gathered at this conclave to discuss the optimal design and time factors involved in the building of a space rocket, the feasibility of artificial satellites, and unresolved questions regarding the application of international law to outer space. The editor took pride in the fact that von Braun, one of the key contributors to the *Collier's* anthology on the moon, had been a major voice at the Stuttgart conference. "The very fact that the word astronautical exists in our language," the text of the editorial noted, "seems proof enough to us that space travel had passed from the realm of conjecture to the field of rather imminent reality."[6]

Yet, at the heart of the series was a powerful undertow of Cold War anxieties. This haunting sense of external danger, even the fear of nuclear annihilation, was expressed on the editorial page of *Collier's* on October 18, 1952. After describing the series themes, the editor concluded that the development of rockets, upon which any future space travel would depend, had been "born of the desire for destruction and conquest in World War II." The Soviet Union—viewed as a country where science had been subordinated to an aggressive political ideology—was actively pursuing its own rocket program. No real international cooperation was possible. *Collier's* called upon the United States to press forward with space exploration, developing all the essential rockets and associated technologies to achieve dominance. In the words of the editor, "the first power that builds and occupies a space satellite will hold the ultimate military power over the Earth."[7]

A key element in the conquest of space was the rocket envisioned by von Braun, a three-stage behemoth with a winged upper stage. On the cover of the March 22, 1952, issue, Chesley Bonestell portrayed the futuristic rocket, firing its second stage as it passes through the upper atmosphere into orbit above Earth. The 265-foot rocket, in many ways, became a signature image for the new space age. Interest in the extraordinary rocket design prompted the Hayden Planetarium to construct a 13-foot scale model under the direction of the planetarium's chief artist and special-effects creator, Walter Favreau. Exhibited in 1952, the scale model, with its special cutaways showing the rocket engines and multiple stages, drew a wide audience, especially young people attracted to the vision of space travel. Von Braun also employed illustrations to capture the special rocket design required for a journey to the moon, a celestial voyage of nearly 240,000 miles. The spaceship, by design, would be huge, a projected 160 feet tall (nine feet higher than the Statue of Liberty), 110 feet in diameter, and powered by 30 rocket engines. At the top would be a sphere housing the crew; beneath the sphere would be two long arms (to serve as cranes) set on a circular track to allow a full 360-degree rotation. The lunar rocket, once assembled at an orbiting space station, would become a spacecraft without streamlining or wings, an aggregation of spheres, tanks, antennae, moving arms, and engine nacelles that could move effortlessly through the vacuum of space. For the long lunar journey, the spacecraft would require huge fuel tanks, carrying 800,000 gallons of liquid propellants. A projected crew of 20 would maintain the spaceship. Solar power, advanced guidance and navigation systems, and radio communications gear would enable the rocket to make the safe round-trip passage to the moon.[8]

Still more imaginative was the *Collier's* piece "Can We Get to Mars?," published in April 1954. Von Braun, in collaboration with Cornelius Ryan, prepared this elaborate blueprint for an expedition to the red planet. A great deal of mystery surrounded the geology and atmosphere of Mars—including the question of life on the planet—so *Collier's* employed the astronomer Fred Whipple to serve as a consultant. As the primary writer for the concluding installment, von Braun himself was intrigued by the idea of a Martian trip. He had worked out the technical details six years earlier (published in a German-language spaceflight periodical). The enormity of the undertaking would give any space visionary doubts concerning its feasibility, even if the technological advances in rocketry and space sciences continued unabated. The trip itself would take eight months, necessitating in von Braun's reckoning a huge flotilla of 10 spaceships. The rockets, to be assembled and launched from an orbiting space station, would be carefully stocked with all essentials for long-term survival, from logistics to fuel to the vehicles for the actual exploration of the Martian surface. The landing on Mars would take place at a polar ice cap, via a "ski-equipped plane." From this base camp, a team of explorers would travel overland to the Martian equator, living in inflatable and pressurized spheres mounted on tractor vehicles.

This fanciful projection of a future Mars expedition did not ignore some of the anticipated problems associated with prolonged weightlessness. Von Braun acknowledged the problem of crew members facing muscle atrophy in outer space. He admitted that no clear remedy was apparent to him or other space visionaries in the 1950s. However, he speculated that this problem—as with the extreme dangers for space travelers from cosmic rays—would be solved in the course of time by means of an exercise regimen or perhaps the creation of "synthetic gravity."⁹

The impact of the *Collier's* series was immense. Never had the prospect of space travel received such systematic and dramatic treatment. Von Braun found himself suddenly catapulted to national fame. In the years that followed, he would emerge as an authoritative spokesman for America's embryonic space program. One concrete measure of von Braun's growing celebrity was an invitation from Walt Disney to assist in the creation of a new TV series on space exploration. Ward Kimball, a producer at Disney, had read the *Collier's* articles with fascination, seeing immediately the potential for a television version. In 1955, the first of three Disney shows, titled "Man in Space," aired nationally. A second program, "Man and the Moon," followed later in the year. The final segment, "Mars and Beyond," aired in 1957. Von Braun served as a key technical consultant for the Disney series, taking pains to assure that every detail was "science factual." The Disney Studios demonstrated great skill in dramatizing future space travel to the moon and Mars. These televised films dovetailed with the opening of Disneyland, where "Tomorrowland" became a popular attraction in the theme park.¹⁰

Popular interest in space also swept the American cinema. A number of remarkable and memorable films expressed the theme of space travel. Robert Heinlein's tale *Destination Moon* was adapted for the silver screen in 1951. Another film, now a classic, was *The Day the Earth Stood Still,* a melodrama about a visitor from a distant world who arrives in a flying saucer on the Mall in Washington, D.C. The visitor, Klaatu, warns earthlings that their violent ways could lead to the destruction of the planet. Other films such as *The War of the Worlds, The Thing,* and *Invasion of the Body Snatchers* portrayed more malevolent and threatening types of alien visitation. This was also the era of the flying saucer sightings, the Roswell incident, and the emergence of the U.S. government's mysterious and top-secret Area 51 test site in Nevada. Television offered young people the adventures of space heroes Captain Video and Tom Corbett. These popular films and programs ran parallel to the more scientifically based projects of *Collier's* and Disney Studios.

In the realm of science fiction, a number of popular writers gained prominence in the decade of the 1950s, using imaginative literature to showcase the compelling themes associated with the new space age. Three men stood out as talented contributors to this genre: Robert Heinlein, Isaac Asimov, and Arthur C. Clarke. Heinlein became a prolific writer of science fiction, attracting a wide audience. A number of youths who read Heinlein in that decade became scientists as adults, some directly involved in the space program. Asimov also emerged in the 1950s as an important science fiction writer with his fanciful tales of clashing civilizations in space. Clarke, an Englishman, worked as a radar specialist during World War II and developed the concept of a geosynchronous communications satellite, but eventually turned to writing. His short story "The Sentinel" became the basis for the 1968 film *2001: A Space Odyssey*.[11]

Whether as a dream or a nightmare, the prospect of humankind exploring outer space became a vital part of American popular culture in the 1950s. Implicit in the fascination with space was a sort of celestial manifest destiny—the shared sensibility that humans would follow the imperative to explore the far reaches of the cosmos. Many shared a keen awareness that the space age, indeed, had arrived, but that human aspirations (and fears) needed to be reconciled with the realities of technology. Arthur Clarke prophesied in 1955 that humans would land on the moon by 1980, give or take 10 years. His prophecy was made before Sputnik and the Apollo program, suggesting the optimism of space visionaries in the decade of the 1950s. As events would prove, reality would outdo fiction, because the space race was just beginning in earnest.

THE SURPRISE ATTACK PANEL

As president, Eisenhower was convinced that sustained intelligence on Soviet capabilities was essential because of the near-complete secrecy surrounding all their military activities. This imperative reflected a persistent fear of a surprise attack on America by the Soviet Union. The potential devastation of such a surprise attack was underscored by the Soviet Union's testing of a thermonuclear device, potentially hundreds of times more powerful than an atomic bomb, in August 1953, just 10 months after the first U.S. hydrogen-bomb test. Such a threat raised the specter of American cities and industrial centers laid to waste. Eisenhower and his generation had experienced directly the attack on Pearl Harbor, which shaped the desire to gain advance warning on the intentions of the Soviet Union. The possibility of a nuclear "bolt from the blue" haunted military planners. Within 18 months of assuming office, Eisenhower established the interagency National Indications Center, chaired by the deputy director of the Central Intelligence Agency (CIA), to prepare "watch reports" and to study strategic warning indicators.[12]

The CIA, along with the Air Force, took an active role in assessing the dimensions of the Soviet threat. Beyond traditional spying to breach the Iron Curtain, the United States utilized a variety of other methods, including airborne photo and electronic reconnaissance missions on the periphery of the Soviet Union. The goal was to record and analyze radar signals and communications traffic to learn the Soviets' capabilities in these areas as well as to photograph air bases, harbors, and other Soviet "targets of interest." Some U.S. spy flights dashed deep into Soviet territory on very dangerous penetrations that sometimes resulted in the loss of American aircraft and their crews to Soviet fighter jets or missiles. Later in the decade, American activities included the construction of a broad band of radar dishes and giant antennas in places such as Alaska, Turkey, and Iran to peer into the Soviet Union to monitor missile tests and communications signals.[13]

These efforts yielded considerable data on the Soviet threat. In the spring of 1954, President Eisenhower decided an additional step was necessary, the creation of a special task force to consider three areas of national security—continental defense, strike forces, and intelligence.[14] The committee was later renamed the Technological Capabilities Panel (TCP). The high-powered group was chaired by James Killian, president of MIT and chairman of Eisenhower's Science Advisory Committee. The TCP's mandate was to assess the Soviet Union's military capability to inflict a surprise nuclear attack on America.

The TCP report, entitled "Meeting the Threat of Surprise Attack," reached Eisenhower in February 1955. The timely recommendations from this document made a profound impact on the American defense establishment in the years that followed. Killian's task force focused on certain vulnerabilities to a potential Soviet attack. For example, the TCP argued that if the United States suffered a devastating nuclear strike, the nation would nonetheless "emerge a battered victor." Given the potential of thermonuclear weapons, the TCP recommended an accelerated program to develop intercontinental ballistic missiles (ICBMs)—Atlas and, later, Titan and Minuteman—and land- and sea-based intermediate-range missiles—Thor, Jupiter, and Polaris.[15]

The TCP report made a forceful conclusion. Although the United States had an offensive advantage, because of its far larger nuclear-armed strategic bomber fleet, it was vulnerable to a surprise attack. The U.S. needed a reliable early warning (radar) system to alert it to an impending Soviet bomber attack, plus targeting information for a counterattack against the U.S.S.R. The TCP recommended a number of specific steps. First among them, the U.S. Air Force's Atlas ICBM development effort should be granted the highest priority. The panel also called for construction of a radar network in the Arctic to detect Russian bombers. Other major suggestions included the design and construction of nuclear-powered submarines, armed with ballistic missiles and capable of remaining submerged for long periods of time.[16]

One section of the TCP report addressed intelligence collection and methods. Its head, Edwin H. Land, inventor of the Polaroid camera, wrote, "We *must* find ways to increase the number of hard facts upon which our intelligence estimates are based."[17] An early form of aerial surveillance was created at the Lockheed Aircraft's famed "Skunk Works" in the mid-1950s. Lockheed's renowned aircraft designer, Clarence L. "Kelly" Johnson, came up with a brilliant design for a new high-altitude reconnaissance aircraft. The spy plane became known as the U-2, and it would play a major role in these tense years of the Cold War. The U-2—once flown in great secrecy— soared above the Soviet heartland with highly sophisticated cameras. The U-2 missions brought back revealing data on Soviet military and rocket activities. The extraordinary cameras for the U-2 were designed by Land and other technicians and could capture images of basketball-sized objects from 70,000 feet above the Earth. What made the U-2 effective for so long was the fact that its cruising altitude was outside the range of Soviet anti-aircraft weapons.

Military planners, though, dreamed about the future use of Earth satellites, which would eliminate the risk of interception and offer greater coverage. For example, if they were launched into a polar (north-to-south) orbit, that would put the entire planet in camera range. The TCP's findings also focused on the Eisenhower administration's need to "legitimize" space reconnaissance, a preliminary step essential to pave the way for the widespread use of satellites. This principle was embodied in the notion of "freedom of space," the right of any nation to orbit a satellite without reference to "air space" limitations. This idea was first discussed in the RAND Corporation's milestone 1950 study. It proposed the use of a small, non-military satellite, placed in an orbit that avoided passage over the U.S.S.R., as a way to establish a precedent for "freedom of space."[18] (RAND started out as a team of scientists and engineers at the Douglas Aircraft Company in 1946, hired by the Army Air Forces to study the possibility of orbiting an Earth satellite. Two years later, it was spun off from Douglas and became a highly influential independent nonprofit think tank.[19]) The TCP reaffirmed the concept, calling for "a re-examination of the principles of freedom of space, particularly in connection with the possibility of launching an artificial satellite into orbit . . . in anticipation of use of larger satellites for intelligence purposes."[20] Any discussion of the concept was welcomed by the Air Force, which had been thinking about such possibilities for years. As far back as in a 1951 study, RAND foresaw that a then-new technique—television— could be harnessed to deliver space-based reconnaissance as an alternative to snapping photos from orbit and then somehow dropping them back to Earth in a timely fashion. That new technique was just coming into millions of American homes in the post-war era.[21]

Other studies further explored the options for recovering images from space, including taking photographs on film and then returning them in a reentry capsule that would be caught in midair as it floated down to Earth. The Air Force moved very quickly following release of the Killian Committee report. Within a month, in March 1955, the Air Force contracted with Lockheed Missiles and Space Company in California to create "a strategic reconnaissance satellite weapons system," designated WS-117L. This became one of America's earliest space programs.[22]

GAINING THE STRATEGIC ADVANTAGE

At the end of World War II, the United States enjoyed a decided advantage over the Soviets in strategic bombers, which at the time offered the sole means of delivering nuclear bombs. The Soviets developed the Tu-4, which was a close copy of the American B-29 Superfortress. However, its ability to threaten the United States was severely limited. Both nations quickly built jet-powered bombers and sought to design ballistic missiles. This rivalry spilled over into the 1950s. To gain further advantage, the U.S. began stationing nuclear-armed bombers in the United Kingdom in the late 1940s. By 1953, it began replacing its piston-engine B-29s and B-50s with significant numbers of a new jet bomber, the very advanced swept-wing B-47, and these bombers were soon dispatched to air bases in Britain, Morocco, and Spain. In addition, the U.S. Strategic Air Command introduced the behemoth B-36, a truly intercontinental bomber that could deliver nuclear attacks on the Soviet Union from bases in the United States.

Soviet Russia faced the potential of a massive nuclear attack in any future war. The Soviets fielded a new generation of jet-powered bombers and other advanced aircraft, but these strides did not guarantee any enhanced security.[23] Joseph Stalin died in March 1953, bequeathing to his successor Nikita Khrushchev the daunting task of resolving the strategic inequality. Khrushchev faced a dilemma: Should he seek to match the long-standing and effective U.S. bomber force? Or, should he reject the idea of building a huge manned-bomber force and opt instead for missiles? The latter option might offer the way to threaten the United States with a large arsenal of intercontinental missiles.[24] In the end, Khrushchev would deploy both bombers and missiles, but he firmly believed in the efficacy of ICBMs, and the Strategic Rocket Forces were created in December 1959.

The United States had started thinking about its own ICBM development as early as 1946, but the project was cancelled after only 15 months, a victim of sharp postwar reductions in military spending.[25] This attitude changed in the wake of Russia's atomic bomb test and the Korean War. In addition to developing the hydrogen bomb, the U.S. resumed intercontinental missile development in the spring of 1951, with a government contract award to San Diego–based Convair to develop the Atlas ICBM.

The task facing the designers was daunting: The Atlas as conceived in 1951 was to fly 5,000 miles and deliver an 8,000-pound atomic warhead. The demand for precision accuracy, though, was still beyond the realm of possibility. The Air Force's bomber standard was to put an A-bomb within 1,500 feet of the target. Military planners believed that an ICBM would have to match that kind of accuracy—after flying thousands of miles from its launch site. That was an enormous problem, and for the moment, at least, an unsolvable one, given the technology of the day.[26] Lacking such pinpoint accuracy, an ICBM with an atomic warhead would simply not be of much use as a strategic weapon—it would land too far from the target to destroy it, even with a nuclear blast. As a result of these concerns, Convair's Atlas received only low levels of funding from its inception in 1951 through 1954. However, a solution to the accuracy issue was on the way in the form of the hydrogen bomb. Its sheer explosive force would obliterate the target even if the weapon missed it by a significant distance.

For both the United States and the Soviet Union, the ICBM offered the optimal delivery system for a thermonuclear device, outclassing existing strategic bombers. ICBMs would supply a swift and lethal response to any nuclear attack and defy defense systems. Both nations began work on so-called "lightweight" hydrogen bombs, meaning the payloads for ballistic missiles could be cut by half or even more. The Atlas design, for instance, evolved from a seven-engine, 160-foot monster to three engines and only 75 feet in length.[27]

A subsequent report by the RAND Corporation concluded that ICBMs had become practical and feasible years sooner than previously expected. The report offered an additional, and crucial, assessment, based on U.S. intelligence about Soviet missile development. The superpower race to field an ICBM was under way, and the Soviets were ahead.[28]

THE REDSTONE MAKES ITS DEBUT

The Pentagon ordered the Army and von Braun at the Redstone Arsenal to use the V-2 experience to develop a tactical rocket that could deliver a nuclear warhead over a range of up to 500 miles, with longer-range ballistic missiles certain to follow. Soon the range was reduced from 500 to 200 miles—a trade-off for a heavier payload—and the missile was given the name Redstone. In developing the new rocket, von Braun used some of the same management techniques successfully employed at Peenemünde in creating the V-2. These included strong internal research and development, in-house production of prototype Redstones, and even test manufacturing of the first few production models before turning that task over to the production contractor, Chrysler Corporation.[29] While this new rocket was based on the V-2, it was, nonetheless, a unique opportunity to design, construct, and test a modern rocket, one that would supersede the pioneering work at Peenemünde in a dramatic way.

When completed, the Redstone represented an important advance over the V-2. The Redstone's warhead and guidance system, for example, was contained in a reentry vehicle that separated from the main body of the rocket (unlike the V-2, where the entire rocket body returned to Earth in one piece). The guidance system used a computer and an inertial navigation system contained in the warhead and relied totally upon onboard instruments. To reduce the missile's weight, the fuel tanks were formed by the outer surface of the rocket rather than being housed separately inside it. Most noteworthy of all, the Redstone was one of the earliest American weapons to combine the atomic bomb and the guided missile, two breakthrough technologies of World War II.[30]

The 69-foot-long Redstone, powered by a liquid-fueled rocket developing 78,000 pounds of thrust, began test flights in August 1953 at Cape Canaveral, Florida. Despite the built-in drawbacks—it took eight hours to assemble, erect, and make ready a Redstone for launch in the field—it was later deployed in Europe, armed with a four-megaton thermonuclear warhead (equivalent to four million tons of TNT).[31] Modified versions of the Redstone would later make major contributions to the American space program, propelling the first American satellite and the first American astronaut into space.

Even as von Braun worked to perfect the Redstone for the Army as a weapon of war, he persisted in his dream that the same rockets could provide the first real steps toward the coming age of human space exploration. Besides his work with *Collier's* and the Disney television shows, whenever he found audiences that would listen, in Alabama or elsewhere, he made his case for space travel.

SERGEI KOROLEV'S NEW ROCKETS

Similarly, in the Soviet Union, Sergei Korolev and others had their own dreams of using the rockets of war for the parallel purpose of advancing the voyage of humans and machines into space. Korolev never surrendered his dreams, and he took every opportunity to discuss his ideas with like-minded colleagues. But the Russian situation was far different: In a closely watched communist society, they were not usually free to advocate space exploration or anything else that the government had not endorsed.

Since the war Korolev and his team had enjoyed some hard-won successes. He labored for some time to convince minister of armaments Dmitriy Ustinov that he required a significant degree of autonomy if his operation, NII-88's Department No. 3, was to achieve major improvements in rocket design, development, manufacture, and reliability. Finally, in April 1950, Ustinov agreed to a reorganization of NII-88 that consolidated several departments under Korolev's leadership to create OKB-1 (Special Design Bureau 1). Korolev was named chief and chief designer of the new bureau. OKB-1's sole mission was long-range ballistic missile development.[32]

Later that year, Korolev's team began launch tests of the R-2, an evolutionary improvement on the V-2 (renamed the R-1 in Soviet production) that doubled the R-1's range to nearly 400 miles. The R-2's design illustrated an interesting aspect of both sides' early ballistic missile development: Though each worked in great secrecy, the United States and the U.S.S.R. nonetheless independently came up with several similar advanced concepts. Both an early American ICBM test vehicle, the MX-774, and the later U.S. Army Redstone shared with the R-2 the decision to dispense with separate internal fuel tanks, using instead the missile's outer skin for fuel storage, thereby achieving significant weight reduction. All three missiles also employed a separate warhead for reentry at the end of the ballistic flight trajectory to avoid problems encountered when the entire missile body reentered the atmosphere.[33]

In addition to missile development, the Soviet Union was moving rapidly in several other areas critical to its Cold War rivalry with the United States. Soviet priorities included development of its own H-bomb, mating that weapon with Korolev's ballistic missiles, countering its rival's overwhelming dominance in nuclear-armed, manned bombers and creating air defense weapons. These advanced technology initiatives were so vital that a new entity was created to manage them. It was called the Ministry of Medium Machine Building (known by the acronym MSM) to mask its true purpose. MSM was headed by Vyacheslav Malyshev, an experienced top manager in the Soviet defense industry.[34]

The R-2 overcame a series of flawed test launches during its development. It saw operational service with the Soviet army and contributed to OKB-1's experience, knowledge, and skills. Most important, it was a vital development on the road to Soviet ICBMs, effectively the final step in the process of learning everything possible from the German rocket experience. Korolev's OKB-1 was ready now to strike out into new frontiers, even as it continued to be influenced by the Germans' achievements.[35]

The R-3, with a projected range of 1,900 miles, was the intended next step in that direction. While a significant improvement over the R-2, its trajectory was well short of the 5,000 miles necessary for a true intercontinental rocket. Korolev therefore proposed to Malyshev that the R-3 be cancelled in favor of development of a full-scale ICBM. According to the noted Soviet space historian Asif Siddiqi, "Korolev almost casually announced to the attendees (at a high-level meeting) that work on the R-3 should be terminated immediately to concentrate forces on going directly to an ICBM."[36]

Malyshev was astonished by the proposal, given his own belief that the R-3 was a critical project for the Soviet military. He rejected Korolev's idea out-of-hand, accusing him of placing his long-term interests in using rockets for space exploration ahead of the country's real needs. Korolev, however, refused to abandon his proposal, whereupon Malyshev turned to threats: "No! Really! He refuses? . . . People are not irreplaceable. Others can be found."[37] In the end, Korolev would get his way. The R-3 was cancelled in 1952, and the Soviet Union moved inexorably toward his goal—the R-7 ICBM.

The demise of the R-3 also paved the way for development of the R-5, which Korolev conceived in order to give Soviet forces a rocket with greater range than the R-2. The R-5 eventually became the first Soviet nuclear-armed rocket, with a range of 750 miles. It employed special thermal shielding to protect its nuclear warhead, which would encounter tremendous heat reentering the atmosphere at a speed of nearly two miles a second. In February 1956, a modified version, the R-5M, successfully carried out the world's first test of a ballistic missile carrying a live atomic warhead.[38] With the success of the R-5M, Korolev received full independence from NII-88 that same year, when OKB-1 became an independent design bureau.

A key aspect of Korolev's success in gaining support for his ICBM was the impression he made on Khrushchev when he briefed the ruling Soviet Politburo on his work not long after Stalin's death. Khrushchev later recalled in his memoirs, "I don't want to exaggerate, but I'd say we gawked at what he showed us as if we were a bunch of sheep seeing a new gate for the first time." Korolev showed the Politburo members one of his rockets, took them on a tour of the launch pad, and attempted to explain how the rocket worked, Khrushchev said. "We were like peasants in a market place," he continued. "We walked around the rocket, touching it, tapping it to see if it was sturdy enough—we did everything but lick it to see how it tasted."[39]

FASHIONING A NUCLEAR THREAT

Less than 10 months after the United States detonated the world's first H-bomb, the Soviets followed suit in August 1953 with their own successful thermonuclear test in Central Asia at the Semipalatinsk site in Kazakhstan. This was a much narrower margin than the several years that separated the two superpowers' tests of their first atomic bombs.

Soviet leaders had originally planned to use an atomic warhead for their planned ICBM, but they eagerly switched to the newly created H-bomb for the long-range missile. Malyshev, the head of MSM and now promoted to deputy chairman of the powerful Soviet Council of Ministers, had both H-bomb and ICBM development in his portfolio. He asked Andrei Sakharov, the lead nuclear physicist behind the Soviet H-bomb, to make an estimate of the weight of a "second generation" H-bomb. Sakharov was uneasy about the request, but decided that he could not refuse; his estimate was five tons.[40]

In October 1953, Malyshev went to see Korolev and his top staff at OKB-1 to discuss their progress in developing an ICBM, soon to be designated as the R-7. He asked Korolev for an estimate of the R-7's payload, or lifting capability, and was not pleased to hear Korolev's estimate of three tons. Malyshev insisted that five tons was the requirement, with no room for negotiation, and his order was soon confirmed at the highest government levels. Korolev and his design bureau were to produce a rocket able to carry a five-ton warhead—several times the capability of missiles at that time—over a 5,000-mile range. This hastily estimated payload requirement of five tons drove the design of the R-7, producing a missile with lifting capabilities that made it useful for launching Russian cosmonauts on space missions for decades to come.[41]

Work on revising the R-7 to carry the increased payload started immediately. The first challenge was how to create the enormous rocket power, or thrust, needed to carry out the missile's intercontinental mission. The solution was found in a pair of engines developed by Valentin Glushko's Gas Dynamics Laboratory starting in 1954: the RD-107 and RD-108. The engineering breakthrough was that each engine—with four RD-107s and one RD-108 powering the R-7—would have not one, but four, combustion chambers. The multichambered concept was a way to avoid expanding a single conventional combustion chamber to the required size. That would significantly increase the chance of creating damaging pressures inside the large chamber that could destroy the engine, a phenomenon known as combustion instability.[42] In addition, the four combustion chambers were fueled by a single turbopump; this resulted in a cumulative thrust significantly greater than using a single combustion chamber. The sharp reduction in the risk of combustion instability was an additional benefit.[43]

The requirement for five main engines in the R-7 led to a missile that looked different from anything seen before. Korolev borrowed a key idea from Mikhail Tikhonravov, a brilliant rocket engineer on his staff. Tikhonravov had long proposed the idea of clustering multiple rocket engines to achieve high payload capabilities. Application of his concept resulted in a revolutionary approach: The single main RD-108 "core" engine would be surrounded by four RD-107 booster engines "strapped on" to the core. Viewed from the side, the boosters made the R-7 much broader at its base than at the top. The strap-ons tapered upward to a point, making it appear almost as if the rocket were wearing a skirt. With all engines burning, the entire ensemble would produce 398 tons of thrust at liftoff, about nine times more than any other Soviet rocket.[44]

At a height of 30 miles, the four strap-on engines fell away from the core stage, which continued in powered flight until its engine cut off and the payload entered into a ballistic trajectory until its reentry into the Earth's atmosphere. (The American Atlas ICBM, used the same layout. All three of the Atlas' engines burned at liftoff and two were later jettisoned, leaving the core, or sustainer engine, to complete the powered portion of its flight.) Korolev's team explored various methods to steer the R-7. They settled on using small steering, or vernier, engines with swiveling nozzles instead of the simple graphite rudders employed on the V-2 and other early rockets. When Glushko refused to manufacture the vernier engines, claiming that they wouldn't be effective, Korolev's staff used a group of young engineers from another scientific institute to successfully develop them.[45]

Korolev also tackled the challenge of developing a guidance system to reliably deliver the R-7's thermonuclear payload on target after a 5,000-mile flight. Because an inertial guidance system of the type used by the V-2 would not provide the needed accuracy, Korolev proposed a combination of inertial guidance and a radio-controlled system to correct deviations from the desired trajectory using the small vernier engines attached to the core engine. In May 1954, the Soviet Union's powerful Council of Ministers issued a decree calling for full development of the R-7. This was followed soon after by Minister of Defense Industries Ustinov's decree that development of the R-7 ICBM was a matter of "state importance."[46]

The May 20 Council of Ministers' decree also set in motion a process that would lead to the development of a new firing range to test the R-7. Besides having outdated facilities, the exising Kapustin Yar site was within range of American radar stations in Turkey, established by U.S. intelligence services to monitor Soviet missile tests.[47] A high-level commission reviewed a number of alternatives and eventually chose a site in Kazakhstan in Central Asia. The Soviet Council of Ministers approved the selection in early 1955.

INTO ORBIT

Another ambition in space was taking hold as well. The idea of an orbiting man-made Earth satellite captured the imagination of Americans and Russians alike. Such a project offered an opportunity to achieve a landmark heretofore regarded as little more than science fiction: shooting a man-made object into space at a speed great enough to place it in orbit around Earth. Both American and Russian experts studying every aspect of the German V-2 immediately following World War II soon concluded that this missile possesed more potential than as a mere weapon. Just days after surrendering to the U.S. Army in the spring of 1945, several Peenemünde engineers briefed members of a U.S. Navy technical team on their rocketry work. They spoke enthusiastically of the possibilities of artificial Earth satellites and even manned space stations.[48]

The Army Air Forces (AAF) agreed to examine the concept, turning again to the RAND Corporation to perform an independent assessment. Even six decades later, the resulting May 1946 report, *Preliminary Design of an Experimental World-Circling Spaceship*, remains fascinating reading. Assuming no major engineering or design breakthroughs, it concluded that a successful satellite launch vehicle *was* possible, perhaps within five years, and with an expenditure of about 150 million dollars.[49] The report observed, prophetically as events turned out, that an Earth satellite, launched into orbit by a 17,000-mile-per-hour rocket, would be "one of the most potent scientific tools of the 20th century." Further, the report asserted, an Earth-orbiting satellite "would inflame the imagination of mankind and would probably produce repercussions . . . comparable to the explosion of the atomic bomb." Nor did the study ignore the many potential military, as well as civilian, uses of Earth satellites. Among those cited were assessments of the weather conditions over enemy territory, post-attack damage assessments of hostile targets, communications relay, and the provision of vastly improved guidance to missiles in flight to their targets.[50] In the end, funding at a level of 150 million dollars represented a huge sum in the post–World War II era of sharply reduced military expenditures, and both the AAF-RAND study and the Navy's work failed to gain the support needed to go forward.

The report noted that the response of the Soviet Union to an American satellite launch, if the United States was first to go into space, was unpredictable. It quoted contemporary Soviet publications asserting that a satellite was an "instrument of blackmail" and that the United States was using "Hitlerite ideas and technicians" in its missile research. The report therefore suggested that the United States place great emphasis on the nonmilitary uses of satellites.[51]

The concept of "freedom of space" would be a driving force for years to come as the United States continued to develop its space policies vis-à-vis the Soviet Union. A "benign" test of the satellite concept, with an experimental satellite on an equatorial orbit that would avoid the Soviet Union, could—if there were no objections from the countries that *were* overflown—establish the precedent of "freedom of space," and thus pave the way for a second "working" satellite, the RAND study asserted.[52]

In the spring of that year, Mikhail Tikhonravov on Korolev's team presented what apparently was the first detailed Soviet assessment of the technical aspects of launching an artificial Earth satellite. The paper proposed using a rocket with multiple engines clustered together to achieve the necessary thrust to orbit a small satellite, which was ultimately used on Russia's R-7 ICBM. Beyond its content, the paper was noteworthy for the fact that Tikhonravov likely was one of the few engineers sharing Korolev's closely held vision of eventually using rocketry for space travel rather than only for weaponry.[53]

While the paper contained no timetable, it seemed to imply that a satellite might be sent into orbit by the mid-1950s, provided the needed resources were made available. The audience's reaction to Tikhonravov's presentation was generally negative, ranging from open hostility and sarcasm to silence. Korolev, however, publicly supported his friend's ideas at the session. Little came of Tikhonravov's efforts at the time, no doubt reflecting the very limited support in Russia for any aspects of rocketry unrelated to the military.[54]

A YEAR TO CELEBRATE

Following the recommendation of the TCP report to fly reconnaissance satellites over the Soviet Union to assess Soviet military activities, the Eisenhower administration in the mid-1950s made a concerted effort to publicly promote the concept of "Open Skies" as a way to reduce international tensions. In late May 1955, just 13 weeks after the TCP report, the president's National Security Council (NSC) took action. In a top-secret document, known as "U.S. Scientific Satellite Program," the NSC endorsed the TCP's recommendation that "intelligence applications warrant an immediate program leading to a very small satellite in orbit around the Earth, and that reexamination should be made of the principles or practices of international law with regard to 'Freedom of Space' from the standpoint of recent advances in weapons technology."[55] Noting that the TCP had specifically suggested such a small satellite, the NSC added: "From a military standpoint, the Joint Chiefs of Staff have stated their belief that intelligence applications strongly warrant the construction of a large satellite. While a small satellite cannot carry surveillance and therefore will have no direct intelligence potential, it does represent a technological step toward the achievement of the large surveillance satellite and will be helpful to this end. . . . Furthermore, a small satellite will provide a test of the principle of 'Freedom of Space.'" The benefits from such a satellite were obvious to the NSC, allowing for continuous surveillance of Soviet installations and the ability to obtain "fine-scale detail" of such objects as airplanes, trains, and buildings on the ground.[56]

The notion of "freedom of space" still remained unsettled in 1955. While long tradition held that nations had sovereignty over the airspace above their territories (the Chicago Convention of 1944 eventually codified such national ownership), the question remained as to just how high such sovereignty extended. Where, if at all, did it end? Where—such as at the height at which a satellite would orbit—would the concept known as "freedom of the seas" take effect in space? That ancient, time-honored tradition held that any nation's ships were entitled to sail the open sea at will, anywhere on the globe.[57]

When the International Geophysical Year (IGY) was announced for 1957-1958, the United States found a convenient "cover" for its plans to launch reconnaissance satellites. The IGY itself was conceived at an unusual meeting in April 1950 in Silver Spring, Maryland, just outside Washington, D.C. The meeting took place at a dinner in the living room of the small brick home of James Van Allen, a physicist then working at the Johns Hopkins University Applied Physics Laboratory. Van Allen, who was to achieve great fame early in the space age for his discovery of two Earth-circling radiation bands, had already acquired considerable experience in American rocket research. He sent Geiger counters aloft to measure cosmic radiation on the first German-built V-2 launched in the United States in April 1946 and continued his scientific work on American-built sounding rockets.[58]

Sydney Chapman, a British geophysicist, had asked Van Allen to host the meeting in Silver Spring, a conclave that would ultimately include eight or ten top scientists to discuss international cooperation in scientific research—what *Time* magazine would later call "a pedigreed bull session."[59] Chapman, Van Allen, and a third eminent physicist, Lloyd Berkner, argued for an international program that would conduct geophysical research of the planet, its atmosphere and space beyond. Berkner was an early advocate of using artificial satellites for scientific research. Van Allen realized that the diminishing U.S. supply of captured German V-2s would exhaust one established vehicle to carry out high-altitude research. A new international scientific initiative—one linked to artificial satellites—would be a reliable way to achieve that end.[60] Walter Sullivan, science editor of the *New York*

Times, later summed up the feelings that drove the dinner participants to dream of new ways to explore the Earth's surroundings. "From the ground, our view into space is hardly more enlightening than the view of the heavens obtained by a lobster on the ocean floor," he wrote.[61]

The suggestion made by Berkner for another global scientific research program in 1957-1958 prompted enthusiastic support at the dinner and gained international scientific support over the next several years. In 1954, the Polar Year was renamed the International Geophysical Year, reflecting a much broader agenda to study the Earth's atmosphere, its oceans, the polar regions, and outer space. Berkner found strong support for the use of Earth satellites to help achieve the IGY agenda. Given how much had to be done, the IGY could not be contained in a mere 12 months; it would run from July 1, 1957, to December 31, 1958.[62]

This was a perfect intersection with the urgent (and highly classified) drive by the U.S. government to establish the principle of freedom of space via the launch of a scientific satellite.

FINDING A ROCKET THAT WORKS

In early 1956, von Braun and his Redstone Arsenal rocket team began work on a critical new project, the Jupiter nuclear-armed intermediate-range ballistic missile (IRBM) with a range of 1,500 miles. This was one of several new weapons systems recommended to President Eisenhower in 1955 by his top-level commission reviewing future American strategic needs. As part of that assignment, which was in addition to ongoing work on the shorter-range Redstone missile, von Braun and his group were placed under a new command, the Army Ballistic Missile Agency (ABMA).

The new boss for ABMA was Army Major General John B. Medaris, a tough up-from-the-ranks officer who had begun his military career as an enlisted man in the Marine Corps. He and von Braun became friends and allies, making common cause in the Army's battle with the Air Force for control of development and operation of medium- and long-range missiles. "Our survival as a rocket-building team is at stake," von Braun said. "Only a tough fighter in command of the ABMA has a chance to keep it alive, and General Medaris is such a man." Years later, Medaris would return the compliment, noting, "Wernher and I were about the most perfect possible match between two men who wished to pursue great projects together." But even as he worked closely with the general and pursued the development of military weapons, von Braun's frustration at being unable to secure government support for his space exploration dreams was unabated. "Galileo, the Wright Brothers, and Thomas Edison wouldn't have a Chinaman's chance here today," he told a journalist in 1956. "They'd be thrown right out of the Pentagon on their ears!"[63]

Von Braun nonetheless found some reasons to believe that the United States would develop a genuine space program, and the IGY was one of them. He believed the Army's Redstone possessed the potential to take center stage in such a space program. His Redstone rocket was doing well in its tests, and he and others inside and outside the military were starting to think about orbiting satellites as a part of the U.S. contribution to the IGY. He therefore was pleased to receive a call in June 1954 from an old friend, Frederick Durant, president of the International Aeronautical Federation (IAF). In an interesting twist in the Cold War setting, Durant, a former naval aviator, worked for the CIA from 1951 to 1954, and continued to provide information to the agency later. The U.S.S.R.'s Academy of Sciences joined the IAF in 1955, and the CIA was most anxious to learn of any comments its representatives made on technical papers presented or Soviet space capabilities.[64]

Durant invited von Braun to his June 25 meeting in Washington, D.C., with Commander George Hoover, chief of the Office of Naval Research (ONR), to discuss placing a satellite into orbit in the near future. Von Braun later recalled that Hoover opened the meeting by saying, "Everybody talks about satellites, then nobody does anything. So maybe we should put to use the hardware we already have." By Hoover's later account, von Braun came prepared with a specific plan to use a modified Redstone. By the end of the session, there was an agreed Army-Navy plan to use a Redstone with lengthened fuel tanks for the booster stage, along with no less than three additional upper stages. The third of these would employ a single solid-fuel rocket to kick a five-pound Navy-developed satellite into an orbit at least 200 miles above the Earth. The time frame for launch was between fall 1956 and November 1957.[65]

The tiny, ultra lightweight satellite would have no room for any instruments, a fact that led to its unofficial nickname, Project Slug, soon changed to Project Orbiter. Launch was to take place at the Equator from a Navy ship specially built to handle large rockets. An equatorial launch to the east would take the greatest possible advantage of the Earth's rotation. This was essential, since the booster-upper stage combination planned in 1954 could not have orbited even five pounds unless it *was* launched from the Equator. Von Braun secured the agreement of his Army boss at that time, Major General Holgar Toftoy, to provide a Redstone rocket as the launch vehicle. Von Braun himself assigned several of his staff members to work on the project, a move accepted by his superiors.[66]

Project Orbiter did not long remain the only candidate to launch America's planned IGY satellite. The U.S. Naval Research Laboratory (NRL), which had reviewed and commented on Project Orbiter, went a step further by seeking acceptance for its *own* satellite-launching rocket. The Navy advocated using its highly successful Viking high-altitude research rocket, which would serve as the first stage of a new rocket; two additional stages would be used to boost a 40-pound satellite into orbit. In other words, the projected weight of the NRL satellite would be eight times the weight of the satellite designed for Project Orbiter. The NRL proposal was submitted to the government in March 1955.[67] In addition to Orbiter and the modified Viking, the Air Force argued for the use of its untested Atlas missile to launch the IGY satellite, but this proposal earned little support. Accordingly, senior Air Force generals viewed the launching of civilian satellites as an unwise diversion from the highest national priority accorded to the development of Atlas as the nation's first ICBM.

The planned IGY satellite, at least in the minds of its creators, was to be a civilian effort. Prestigious organizations such as the National Science Foundation, an independent federal agency, and the National Academy of Sciences endorsed the civilian nature of the project. Nonetheless, it was understood that the military alone possessed the resources and experience necessary to develop and launch rockets and satellites. Accordingly, responsibility for the launch vehicle selection was assigned to the Department of Defense.[68]

On July 28, the White House publicly announced its strategy to protect the nascent U.S. reconnaissance satellite program by emphasizing the peaceful use of space for international scientific purposes. Eisenhower's press secretary James Hagerty observed: "The president has approved plans by this country for going ahead with the launching of small, Earth-circling satellites as part of the United States participation in the International Geophysical Year. . . . The president expressed personal gratification that the American program will provide scientists of all nations this important and unique opportunity for the advancement of science." A possible reason for the timing was American concerns that the U.S.S.R. might move first to announce its own IGY satellite plans. As early as January of that year, Radio Moscow had reported that Soviet experts believed they could send an Earth satellite into orbit in the near future.[69]

Durant announced the U.S. plan at the IAF congress in Copenhagen, Denmark. Academician Leonid Sedov, chairman of the Commission for Interplanetary Communication of the U.S.S.R. Academy of Sciences, also attended the meeting. Soon after Durant spoke, Sedov announced the Soviet Union's own commitment to launch a satellite. "From a technical point of view, it is possible to create a satellite of larger dimensions than that reported in the newspapers which we had the opportunity of scanning today," he said. Sedov then added, "In my opinion, it will be possible to launch an artificial Earth satellite within the next two years. The realization of the Soviet project can be expected in the near future."[70]

The reference to a Soviet satellite of "larger dimensions" than the one announced by the United States came more than two years before the launch of Sputnik 1. Was it perhaps intended as a tantalizing tease? At the time Sedov spoke, the Soviets had no plans for launching a satellite. Nor would Sedov likely have had much contact with Korolev's design bureau. Nonetheless, he clearly knew something was being planned; it is highly improbable that he made the comments without authorization. Regardless of the circumstances behind Sedov's comments, they and the Eisenhower administration's announcement together generated significant press attention.[71]

Back in Washington, on August 4, 1955, an advisory group named the Stewart Committee—after its chairman, Homer J. Stewart of California Institute of Technology's famed Jet Propulsion Laboratory—issued a formal report endorsing the U.S. Navy's entry, which soon took the name Vanguard. The Army was predictably opposed to the decision, citing its belief that Redstone could orbit a satellite sooner than Vanguard. The Army had a proven flight-tested rocket, while Vanguard existed only on paper. Senior Army officials also noted the missile was set to enter production shortly.[72]

The reasons for this decision have remained controversial. One factor that may have shaped it was the fact that the Army's team was composed almost entirely of Peenemünde veterans, and the Redstone owed much of its design lineage to the V-2. Vanguard, by contrast, was a completely made-in-the-U.S.A. entry.[73] Also influencing the selection may have been the desire to select as the IGY satellite launcher the rocket that came closest to having a civilian pedigree. On that score, Vanguard was the clear winner. Orbiting of a benign scientific satellite as part of America's IGY contribution was in keeping with the overarching goal of establishing freedom of space as quickly as possible. That, in turn, would pave the way for the WS-117L strategic reconnaissance satellite to fly over the Soviet Union without credible legal challenge.[74]

Von Braun, of course, was disappointed by the Stewart Committee's decision. "This is not a design contest," he said. "It is a contest to get a satellite into orbit, and we (the Army) are way ahead on this." The most he could secure, however, was a promise that the Redstone would be a backup for Vanguard. With the unofficial sanction of his superiors at the ABMA, von Braun and his team continued to develop a satellite and booster rocket, anticipating a time when it would be needed. "We bootlegged work on the satellite. Night after night, men . . . put in endless hours on their own," he said.[75]

The Navy's Vanguard development effort did not get off to a strong start. Almost immediately after development work began, it became clear that the Department of Defense, focused on a huge build-up of missiles and bombers to counter the perceived Soviet threat, did not regard Vanguard as a high priority. That reality was underlined in October 1955, when the Vanguard's builder, the Glenn L. Martin Company, was selected by the Air Force to build America's second ICBM, Titan, as a backup to the Atlas. Realizing the Titan would be a much larger and more profitable program than Vanguard, Martin transferred many of the experienced engineers expected to work on Vanguard from Baltimore to Denver, Colorado, to work on the Titan.[76]

Vanguard had other problems as well. These included a redesigned first stage that virtually eliminated all traces of the Viking research rocket on which it was supposed to be based and replacing the original Viking engine with a new one built by General Electric. Looming over all the project's difficulties was the specter of an out-of-control budget. What started out as a reasonably priced 20-million-dollar program soon doubled, and then doubled again, to just over 80 million dollars. In April 1957, Eisenhower's Bureau of the Budget director informed him that the total cost estimate had risen to nearly 110 million dollars. The president was not happy to hear this news; he directed that every possible effort be made to avoid "gold-plating" and control costs. (The Central Intelligence Agency provided 2.5 million dollars to the Vanguard project, its "civilian" nature notwithstanding, likely looking ahead to the time when it would launch reconnaissance satellites.[77])

Yet another problem soon emerged to dog Vanguard. Although the contract with Martin called for the orbiting of a 21.5-pound payload, or satellite, delays in testing resulted in an NRL decision to sharply reduce the required payload for the Vanguard's first satellite orbit attempt. To ensure that Vanguard would orbit at least one satellite during the July 1957 to December 1958 IGY, NRL opted to use a miniature 6.5-inch sphere, weighing under four pounds, and intended as a test satellite, for its first try to orbit.[78]

Even as they kept their satellite plans and dreams alive, von Braun and his team remained focused on the Redstone Arsenal's major project, the Jupiter IRBM. The work on that missile would lead to the Army eventually winning the satellite-launching contest. One of the most daunting problems facing the new long-range ballistic missiles then under development was the need for a reentry vehicle (RV) that would safely carry the missile's thermonuclear warhead back through the Earth's atmosphere at the end of its flight. The reentry took place at very high speeds, generating an enormous heat buildup. It was vital to test the Army's RV design in space before it was tried out on the Jupiter.[79]

Von Braun recognized that the Redstone modifications made for Project Orbiter also made the missile ideal for testing RVs. As von Braun's colleague Ernst Stuhlinger later wrote: "It is not known which of the two concepts emerged first in von Braun's thinking: Redstone as a reentry test vehicle or Redstone as a satellite launcher. Both applications of a launch vehicle had been obvious to him for years. When the opportunity arose, he was prepared. Either application would have enabled the other one."[80]

Twelve Redstones were modified for use in the RV test program, using the designation Jupiter-C, though in truth they bore no similarity to the Jupiter immediate-range ballistic missile. They were, rather, Redstone missiles lengthened by a few feet to carry more propellant, equipped with upper stages similar to the Project Orbiter design, and modified in other ways to carry out the RV missions—and, not incidentally, to create an ideal satellite launcher. Von Braun noted later, "We did *not* point out that, with minor modifications, they [the Jupiter-Cs] could also serve as satellite launch vehicles."[81]

On September 20, 1956, the first Jupiter-C stood on its pad at the U.S. Air Force's Cape Canaveral launch facility, ready to fly. It was configured with three live stages and a dummy, or inert, fourth stage. Von Braun, of course, was not alone in realizing the Jupiter-C's potential, with a functional fourth stage, to kick a payload into orbit. As he oversaw launch preparations from his office, a phone call came from his ABMA boss, General Medaris, who advised him firmly: "Wernher, I must put you under orders personally to inspect that fourth stage to make sure it is not live." Pentagon officials, concerned that von Braun might harbor an irresistible desire to beat Vanguard to orbit, had ordered Medaris to ensure that didn't happen.[82]

Though an orbital mission was not possible on this flight, the top (fourth) stage reached an altitude of 682 miles and a range of 3,355 miles, both records, clearly demonstrating the American potential to launch a satellite. Nonetheless, another 16 months would elapse before another, nearly identical Jupiter-C would have its chance to perform that mission.

THE LOOMING THREAT

In 1957, the Soviet Union—under a shroud of strict secrecy—began to take concrete steps to launch the first Earth satellite. This would be a dramatic contribution to the IGY by the Communist state—and one with vast propaganda value. The impetus for the endeavor stemmed from the United States' July 1955 announcement of its own plan to launch a satellite within the time frame of the IGY. This provided the Soviet Union with a clear warning of American intentions. Now the task at hand for Korolev and his team was to beat the Americans into space. The R-7 missile, already at hand, became the chosen carrier for launching the Soviet satellite into orbit. The nose cone of the R-7 was redesigned to carry a satellite rather than a thermonuclear device. Faced with the urgent task to leap ahead of the Americans, the Korolev team quickly built a simple and lightweight satellite—*Prostreishiy sputnik,* or "PS."[83]

Korolev moved fast to test a reconfigured R-7 as the launch vehicle for the Soviet satellite. He was anxious to have the satellite in orbit before the beginning of the IGY. The first three launches of the R-7, in May-July, were failures, creating fears that the United States might indeed be the first with a satellite in orbit. When a fourth R-7 launch proved successful on August 21, Korolev was elated. The missile followed a 4,000-mile trajectory to the Kamchatka Peninsula in the Soviet Far East. On September 7, Korolev repeated the R-7 test flight. Nikita Khrushchev, who witnessed the September firing of the R-7, gave his go-ahead for a satellite launch, seeing clearly the propaganda potential for the Soviet Union on the eve of the IGY celebrations.

Still fearing an imminent American satellite launch, Korolev decided to act decisively. A powerful R-7 booster was transported to the launch pad at Baikonur on the morning of October 3, 1957, escorted on foot by Korolev and members of his team. Technicians spent the following day and night getting the rocket ready for a momentous launch.[84] The space age was about to be born.

TAKE THE
A-ROCKET

In the years after World War II, atomic power seemed like the logical future of space rocketry. The staggering magnitude of nuclear energy had been demonstrated by atom bombs, and nuclear reactors promised to channel this power for peaceful uses. Halfway through a century that had seen technological leaps from the horse to the jet aircraft, it seemed reasonable to suppose that the atom would soon serve humankind as obediently as radio waves and the internal-combustion engine. Military designers drew up concepts for atomic tanks and actually put considerable development into designing an atomic-powered airplane. Detroit created a study model of the Nucleon, the personal atomic automobile. In this climate, a reactor-powered rocket didn't seem outside the realm of possibility.

Leading space writer Willy Ley wrote of these atomic rockets in his 1949 book, *The Conquest of Space*.[85] America's greatest space artist, Chesley Bonestell, combined the expectation of near-future nuclear rocket power with the iconic image of Wernher von

Artist's conception of a nuclear-powered spacecraft in orbit over Mars.

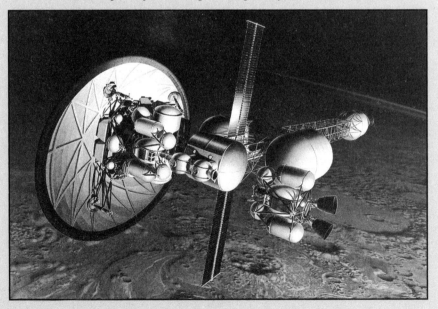

Braun's V-2 to produce realistic paintings for books and popular magazines of the time. He depicted transcontinental commercial atomic rocket planes and the moon landing of a towering, single-stage atomic space rocket so powerful that it could fly all the way to the moon and back.

Just as nuclear reactors turned out to be more difficult to manage than first anticipated, rocket development proved so expensive that "incremental advance" became the standard practice, especially under pressures of budget and time. Improving past designs little by little wherever possible tended to be far more reliable than venturing a design that was completely new. The V-2 precedent put Soviet and American engineers alike on the path to improving and enlarging upon its liquid-fuel concept rather than replacing it entirely with a reactor. The whole space race was accomplished on both sides with conventional liquid-fueled rockets, from the Sputnik's R-7 to the mighty Apollo-Saturn V.

Wernher von Braun nonetheless saw a real future for the atomic engines. NASA had even put its own development into the idea under the NERVA program: Nuclear Engines for Rocket Vehicle Application. Theoretically, a nuclear rocket would be more efficient than a conventional rocket, making it ideal for "long haul" flights such as interplanetary missions to Mars. A reactor would superheat its liquid fuel, rather than burning it, and would be fired only in space, leaving no radiation in Earth's atmosphere. The nuclear engine could be refueled in space and reused for years. Such an asset would drastically reduce the cost of space-flight operations. Von Braun had artists draw up the vital role this nuclear rocket would play in the post-Apollo space program.[86] His NERVA "space tug," once launched, would ferry loads back and forth from the moon to high Earth orbit and help carry astronauts to Mars. NERVA engines were built, and in the remote desert of Nevada, tests proved out the concept. We might have seen nuclear rockets after all, but NERVA was cancelled along with von Braun's other visions under President Richard Nixon, who put an end to the Apollo lineage in favor of the space shuttle program.

Since the NERVA program, British-born American physicist and mathematician Freeman Dyson seriously proposed the idea of a rocket propelled by detonating atomic bombs for thrust, under the name Project Orion. Today such ideas, like all nuclear rocket engines, exist only in science fiction. ∎

SPUTNIK NIGHTS

n late September 1957, Sergei Korolev abandoned OKB-1, his experimental design bureau located in the suburbs of Moscow, for a new top-secret rocketry complex in Kazakhstan. The move was abrupt and unannounced, shrouded in mystery as all of Korolev's activities were. As a leading designer of long-range ballistic missiles, he remained largely unknown in his own country, notwithstanding the fact that he would soon be at the cutting edge of Soviet-designed robotic and manned spacecraft. The previous month, he had overseen the successful launch of the R-7 (SS-6) rocket, the Semyorka, a missile with the potential to hit distant targets with nuclear weapons. Korolev's shadowy move to Kazakhstan signaled that the R-7 was now part of a clandestine scheme to launch the world's first artificial satellite. The planned launch, keyed to fall in the International Geophysical Year, was a risky undertaking, given all the technical uncertainties. But the bold endeavor promised dramatic propaganda possibilities—an occasion to showcase the latest strides in Soviet technology and to beat

OPPOSITE: *Sputnik 2, the second artificial satellite, flies over the United States, November 3, 1957.*

the Americans to the punch in what would become one of the most memorable technological milestones in human history.[1]

Taking leave of Moscow, Korolev flew with his new satellite, christened *Sputnik* ("fellow traveler"), to a distant test range in Central Asia called Baikonur. Tight secrecy surrounded the entire project. He aimed to mount the satellite in the nose cone of a powerful R-7 for an early October launch.

Reaching the threshold of the historic launch had come only after a long process of preparation. The previous month had been devoted to the frenetic drive to complete the construction of the satellite. Hard-driving and alert to the smallest details, Korolev had pushed his staff relentlessly to fashion the spherical-shaped satellite into what he later described as "a simple and expressive form, close to the shape of natural celestial bodies."[2] The resulting metallic sphere was highly polished, at Korolev's insistence, to ensure that the satellite properly reflected the sun's rays—both to reduce heat and to increase the probability of being seen while in flight above Earth's atmosphere.

Mikhail Khomyakov had served as the primary designer for the satellite, and Korolev's talented workers at OKB-1 had stamped the sphere in two halves, which were then vacuum-sealed. The small satellite contained a radio transmitter, batteries, and temperature-measuring instruments, weighing 184 pounds—for the time, a massive object to lift into space. On the eve of the flight, Korolev had his creation placed on a special pedestal and draped in velvet, giving the object the sort of reverence typically accorded to a Fabergé egg. That was his way of expressing to his workers the unique importance of Sputnik. This small, shiny object was the harbinger for a new stage in the history of rocketry. Upon first seeing Sputnik, famed Russian test pilot Mark Gallai expressed a certain deference and amazement, calling the satellite an "elegant ball . . . with an antenna thrown back like a galloping horse."[3]

Once Sputnik reached remote Baikonur, the process of rocket assembly began in earnest. The satellite was placed carefully in the nose cone of the R-7, a task completed in a special "space room." Here a severe regime of cleanliness prevailed, with technicians in white smocks attending to the many details of the launch preparation. Mikhail Rebrov remembered the thoroughness of Korolev, who was known to those in the emerging Soviet space program simply as the "Chief Designer." Whenever Korolev appeared in the hangar, everyone fell silent, being keenly alert to his demanding standards, unforgiving posture toward failure, and the high purpose associated with the top-secret Sputnik project.[4]

The design of Sputnik had no historical parallel; it was an original piece of technology, built with studied simplicity and governed by one requirement—the satellite could not weigh more than 220 pounds. The spherical shape met Korolev's demand for an object that could fit easily into the interior of the R-7 nose cone. The rocket design required the easy release of the satellite once the fairing on the nose cone separated. Instrumentation reflected Korolev's austere approach: He decided to have two radio transmitters, powered by silver-zinc batteries, one with a frequency of 20,005 megahertz, the other at 40,002 megahertz. The choice of these frequencies meant that radio signals from Sputnik could be detected at extreme distances, in particular by amateur radio operators tuning in the shortwave and ultra-shortwave ranges. The radio signals from Sputnik would be emitted in the form of telegraphic pulses lasting approximately 0.3 seconds. When one radio transmitter was operating, the other was in the pause mode. By plan, the radio signals sent from Sputnik, the staccato sound of "beep, beep, beep," would entertain a global audience of trackers, amateur and scientific observers alike, for a period of approximately two weeks.[5] The Sputnik satellite was small, measuring a mere 22 inches in diameter, but Korolev remained confident that its highly reflective polished surface would make it clearly visible to the naked eye. The Soviets wanted the world's first artificial satellite to be tracked—by sight or sound—because this observation by a worldwide audience would confirm a new benchmark in technology.

Sputnik rode atop the R-7, the most powerful operational rocket in the Soviet inventory. The imposing missile consisted of a core rocket surrounded by four tapered boosters. At liftoff the core and boosters ignited simultaneously. Each of the strap-on booster rockets was fitted with one engine, producing approximately 110 tons of thrust at sea level. About two minutes after liftoff, the four strap-on boosters separated from the core. The rocket would continue to fire, lifting the R-7 to altitude and inserting the satellite into orbit. The development of the R-7 reflected the impressive Soviet capacity to lift huge payloads into orbit (and correspondingly the potential to send missiles armed with nuclear warheads to strike distant targets). Used in 1957-1958, the R-7 would place into orbit several artificial satellites. Its successor, the Vostok ("east") launch vehicle, first introduced in September 1958, was an adaptation of the R-7 design, capable of lifting payloads of five tons into orbit. This rocket was the one used to place the first Soviet cosmonaut into space in 1961.[6]

The dramatic launch of Sputnik into orbit took place on Friday, October 4—at exactly 22:28 hours, 34 seconds (Moscow time). The fateful day had been filled with minor delays. Yet all of these frustrations and the many weeks of arduous work faded from consciousness as the R-7 lifted from the ground with a thunderous roar. In the command blockhouse, Korolev observed the historic launch through periscopes with a number of key associates: Vasiliy M. Ryabikov, Valentin P. Glushko, Mstislav V. Keldysh, and Nikolay A. Pilyugin. The man who had been delegated the responsibility of launching the rocket was Boris S. Chekunov. At the instant Chekunov pushed the switch, the five engines on the R-7 ignited into a bright fiery conflagration, and the ground shook under the impact of 1,120,000 pounds of thrust.[7] The rocket rose majestically into the sky, accelerating to 17,400 miles per hour. After reaching an altitude of 142 miles, the core engine shut down. The satellite with its four spring-loaded antennae then entered into a free orbit around the Earth. The orbit inclined at an angle of 65 degrees to the plane of the Equator, following an elliptical path ranging from 141.7 miles to 588 miles from Earth. No sooner had Sputnik begun its orbital passage than, on cue, radio signals beamed downward; the Kamchatka station in the Russian Far East became the first to confirm the reception of the distinctive "beep, beep, beep" signals from the orbiting Sputnik. Nikita Khrushchev, attending a conference in the Soviet Ukraine, received the news of the successful launch at approximately 23:00 hours. Khrushchev shared Korolev's enthusiasm, announcing that a new era of missiles had arrived, one that "demonstrated the advantages of socialism in actual practice."[8]

First word of Sputnik reached Washington in a dramatic way, descending, as it were, on the IGY conference at the National Academy of Sciences building like a bolt of lightning. Walter Sullivan, the science reporter for the *New York Times,* had come to the nation's capital to attend the IGY meetings. Part of the buzz surrounding the Washington conclave was the imminent launch by either the Soviet Union or the United States. Inside the ornate domed interior of the National Academy of Sciences building on Constitution Avenue, a full-scale model of the Vanguard satellite had been placed on exhibit. The Vanguard mock-up, with the satellite buzzing around the Earth, had caught the attention of Sullivan and the attendees, including the Soviet representatives. The latter shared no details on the announced plans of the Soviet Union to launch its own satellite. On the night of October 4, Sullivan attended a special party at the Soviet Embassy on 15th Street, joining a number of prominent American scientists, including James Van Allen from the University of Iowa; William Pickering, the director of the Jet Propulsion Laboratory in Pasadena; Herb Friedman, a research scientist with the Navy rocket program; and other luminaries. Also attending the party was Anatoliy Blagonravov, the leader of the Soviet delegation to the IGY conference. Blagonravov—though a high-ranking official and privy to many secrets concerning the Soviet rocket program—was unaware of the firing of a rocket from Baikonur earlier in the day.

Shortly after Sullivan arrived at the embassy, the Soviet attaché pulled him aside for a phone call. In the breathless conversation that followed, the *New York Times* news desk informed Sullivan that they had just received a wire report from Reuters, announcing that the Russians had placed a satellite in orbit; the satellite, named Sputnik, at that very moment was circling the Earth and beeping radio signals to tracking stations below. When the excited Sullivan returned to the party, he was keenly aware that he alone possessed this stupendous news. Without hesitation, he hurriedly informed the senior American scientist in attendance, Lloyd Berkner, of the remarkable Russian space feat. Berkner, in turn, lost no time in sharing the news. He clapped his hands to get the attention of the assembled partygoers and then announced the news, offering a toast to the Soviets for their success in launching the world's first artificial satellite. The stunned onlookers joined Berkner in the toast. The Soviet delegates, as Sullivan later reported, were ecstatic, smiling "like they had swallowed a thousand canaries."[9]

LOOK TO THE SKIES!

The spectacle of Sputnik in orbit above the atmosphere—with the relentless "beep, beep, beep" of its radio transmissions—quickly caught the attention of a global audience, one that included scientists, amateur ham radio operators, military intelligence operatives, and countless spectators drawn to rooftops with their binoculars. The event was a human milestone with no historical precedent and, for the Russians, a technical triumph of immeasurable value. On October 7, *Le Figaro* caught the mood of the moment, proclaiming in a banner headline "Myth has become reality: Earth's gravity conquered." With thinly disguised delight, the French newspaper observed that the United States—a nation rarely humiliated in the "technical domain"—now found itself caught up in a spiral of "delusion and bitter reflection."[10]

On that same day, Britain's *Manchester Guardian* expressed the view that the Soviet Union now possessed enormous technical prowess, a newly acquired potential to send rockets to the planet Mars. On a more apocalyptic note, the *Guardian* noted the threatening military specter of Sputnik, opining that the "Russians can now build ballistic missiles capable of hitting any chosen target anywhere in the world."[11] Having launched an artificial satellite during the IGY, the *Guardian* noted, affirmed certain peaceful intentions on the part of the Soviet Union, but the West would be foolhardy to ignore the strategic implications of Sputnik in the context of the Cold War.

A week later, *Time* magazine made an effort to define the epochal meaning of Sputnik for Americans. The magazine interpreted the whole episode as another chapter in the long history of human exploration, the quest for new frontiers: "The launching of the Russian satellite is man's first successful attempt to navigate the ocean of space around the Earth." Coming as a stunning surprise, the Soviet technical feat, in the words of *Time*, represented a "Red Triumph." America's rival satellite, Vanguard—at the time of publication still mired on the ground—could not match Russia's orbiting satellite in the critical categories of "weight, orbit, and altitude." Sputnik weighed an astounding 184 pounds. By contrast, the Americans hoped to put into orbit a test version of its Vanguard satellite, weighing less than four pounds.

The Russians had impressed the editors of *Time* with their boldness, catapulting Sputnik into an elliptical orbit that would pass over nearly all of the inhabited Earth (the American Vanguard projected an orbit passing south of Europe and clear of most of the Soviet Union). In addition, Sputnik's apogee in orbit was 559 miles, by the magazine's calculation well outside the outer fringes of the atmosphere. For *Time*, the reason for the American defeat "in the race for space" had become apparent in the first week of October: The United States had not made use of its larger military rockets, depending instead on the Navy's Viking—an underpowered rocket, in the words of the magazine "barely capable" of launching the small Vanguard satellite into orbit. This critique of Vanguard, in a matter of weeks, proved to be prescient. Taking the longer perspective, *Time* concluded that Russia's scientific community had matured, and the dependence on German specialists had long passed—"The Russians are now on their own."[12]

Sputnik would orbit the Earth until late January 1958, when it fell from orbit and made a fiery descent into the Earth's atmosphere. At the start, Sputnik took 96 minutes and 12 seconds to complete an orbit of Earth. But with the passing of time, atmospheric resistance began to shorten the time frame for its path around the Earth, with the orbital dimensions being reduced at a steady and inexorable rate. The carrier rocket, initially in approximately the same orbit, encountered the same forces. Both Sputnik and its launch vehicle were destined to only a short life span. The steady erosion of Sputnik's orbit, however, was not evident in the first weeks of October as the satellite made its dramatic passes around the planet.

The first appearance of Sputnik over a major city, heralded by its radio signals, became a moment for celebration. The Soviet Union took elaborate steps to alert the world to its ever-shifting orbital path. On day two, at a moment of intense global curiosity, Sputnik appeared over Berlin no less than 13 times. Dublin, to the surprise and delight of locals, was graced with the same number of visitations. Other cities encountered fewer appearances of Sputnik on that memorable day: New York, 7 times; London, 6; Tokyo, 6; and Washington, 5. The orbits of Sputnik could be tracked visually only at night, however. Seeing Sputnik with the naked eye as it moved swiftly across the night sky—often with the brilliance of a star—left a profound impression on all observers. No human object had ever gained such a stupendous altitude and orbited the Earth in such a dramatic fashion. Observers were awestruck by Sputnik, especially since many could track its orbits with the naked eye. In reality, as reported later, these observers were often tracking the more reflective carrier rocket, not necessarily the smaller Sputnik satellite.[13]

The Russians quickly moved to augment their stunning October surprise by launching a second satellite—named simply Sputnik 2. Always alert to anniversaries, Khrushchev—the force behind the decision—ordered the launch of Sputnik 2 to take place on the eve of the 40th anniversary of the Bolshevik Revolution on November 7. The official order to make the launch, issued on October 12, reflected Khrushchev's wish to exploit fully the immense propaganda triumph that had come with the first Sputnik. However, preparations for a second launch proved to be a daunting undertaking for Korolev and his team, who were forced to work in a narrow window of time—less than a month. The task became more complex with the decision that Sputnik 2 would carry a dog into outer space. For the historic launch, a mixed-breed female terrier named Laika was selected—and Laika's inevitable martyrdom would carve out a new chapter in the chronicles of space history. The Soviets had conducted a number of high-altitude rocket launches with dogs, but in the case of Laika, there was no plan for the canine's safe return. A significant amount of improvisation accompanied the creation of Sputnik 2, since the upper-stage satellite had to be manufactured without reference to an existing design. The engineers made basic drawings and these were handed directly to workers in the shops. The final assembly was completed with what Boris Chertok called "on-the-spot fitting."[14] One key change in the design of the Sputnik 2, however, was the decision not to separate the satellite from the core booster upon reaching altitude. Consequently, the second satellite with Laika on board consisted of the entire R-7 booster.

The launch of Sputnik 2 occurred on November 3; the liftoff took place without mishap, propelling the satellite with its canine passenger into orbit. The electrical power on board was finite, enough to sustain the instrumentation and the life-support system for Laika for approximately six days. The dog was placed in a pressurized capsule, equipped with water and food dispensers. Electrodes measured Laika's pulse and blood pressure. Telemetry from Sputnik 2 offered a partial chronicle of Laika's life in space: She appeared to survive the high-g launch without difficulty and settled into a relaxed posture for the duration of the flight. Reportedly she barked and ate some of her food on her ill-fated trip into outer space. Once the electrical power died, Laika succumbed to the excessive heat in the capsule. Later, the Soviets promoted a new cigarette brand with Laika's image on the package to honor the canine pioneer.[15]

Unlike Sputnik 1, the second satellite was fitted with a number of scientific instruments, a concerted move to make the second launch more than an aerial stunt—and to link it to the goals of the IGY. Sputnik 2 entered an orbital path with an apogee of 1,038 miles. Western scientists were amazed at the extent and sophistication of the instrumentation on this second artificial satellite. The Soviets made a number of precise and detailed measurements of solar radiation, among other experiments, on Sputnik 2. The final payload for the satellite was slightly over 1,100 pounds, which was a quantum leap from the weight of Sputnik 1. The implication for Western military planners was obvious and threatening: The Soviets with the R-7 rocket possessed the means to design a genuine intercontinental ballistic missile.[16]

The rapid sequence of two Sputnik launches during October-November 1957 heralded the advent of a new space age. In this new arena for human exploration, the Soviet Union cast itself as the pathfinder. Moscow took pains in its official media to portray the stunning successes of their space program as evidence of the superiority of Russia's socialist system. Marxism-Leninism had opened a new and bright future for humankind with its commitment to science and human progress. Little reference, if any, was made to the underlying importance of rocketry to its national security—the urgent need to build an effective intercontinental ballistic missile.

Furthermore, Moscow was planning additional space spectaculars in the near term and beyond: Sputnik 3 would be launched on May 15, 1958, to be followed by Sputnik 4 and Sputnik 5 two years later. The Sputnik series would culminate in 1961 with Yuri Gagarin becoming the first human to orbit Earth. As it turned out, this was the golden age for the Soviet space program, a sequence of successful launches made possible by the powerful R-7, a military rocket that was cleverly deployed by Khrushchev for use in the emerging Soviet space program.

FACING THE RUSSIAN CHALLENGE

As long as the world of Soviet rocketry remained concealed behind a thick curtain of secrecy, Americans were denied any balanced understanding of the Soviet Union's actual potential in what was now emerging as a space race. Most Americans came to the grim realization that these Sputnik launches represented a milestone: No longer could the Americans dismiss their Cold War rivals as technological inferiors. The Soviet Union had emerged as a superpower with a coherent program for space exploration. Some Americans understood that these space triumphs also portended ominous military implications for U.S. national security, despite the fact that they had been pursued under the cloak of scientific research.

The popular response to Sputnik 1 and Sputnik 2 is best described as an awkward mix of incredulity, a curiosity about Soviet intentions, and a growing anxiety over the unprecedented national humiliation. Before the twin launches, Americans had little doubt that the IGY Earth satellite honors indeed would belong to the United States. One stunning example of this prevailing attitude was a book published earlier that year. It was written by Martin Caidin, considered one of the country's leading and highly respected writers on aviation and aeronautics. Caidin also wrote fiction, most notably the novel that became the basis for *The Six Million Dollar Man* television series. Unfortunately, as events turned out, *Vanguard! The Story of the First Man-Made Satellite* was published as nonfiction. Following the story line of his title, Caidin began: "One day during the International Geophysical Year of 1957-1958, the entire world will focus its attention on a desolate stretch of sand lying along the . . . coast of Florida. On a date which is yet to be announced, a small scientific and military army will concentrate at Cape Canaveral . . . the heavily-guarded and secret launch site. . . . They will come to this lonely part of Florida to report to the Earth's people one of the greatest moments in their history: the launching of Vanguard, the first artificial space satellite."[17]

The mood of national concern quickly took on a political aspect. President Eisenhower, through his scientific adviser James R. Killian, Jr., appointed in the immediate aftermath of the Sputnik launch, assured an alarmed electorate that the Soviet satellite was not consequential.[18] Lyndon B. Johnson, then the Democratic Senate Majority Leader, took exception to the reassuring words uttered by the Eisenhower administration, arguing that Sputnik amounted to a latter-day Pearl Harbor. Once evoked, this sense of falling behind the Soviets in rocket technology would eventually evolve into the argument in the 1960 presidential election that America faced a "missile gap."

The autumn of 1957 became a season for renewed debate over national character and priorities. The postwar era of prosperity and consumer culture was viewed by many as hedonistic, a social context where Americans had lost their way. In contrast to this perceived decline in values, the Russians emerged in the minds of many Americans as formidable—highly disciplined, dedicated to science, intent on dominating the world. Edward R. Murrow, the legendary CBS news reporter, argued that Sputnik had "shattered a myth," the notion that scientific advances were not possible in a Communist dictatorship. "We failed," Murrow argued, "to recognize that a totalitarian state can establish its priorities, define its objectives, allocate its money, deny its people automobiles, television sets, and all kinds of comforting gadgets in order to achieve a national goal. The Russians have done this with the intercontinental missile, and now with the Earth satellite." Edward Teller, a nuclear physicist attuned to the Cold War tensions and a fervent anti-communist, echoed Lyndon Johnson's position that the United States had just been defeated in a battle more important than Pearl Harbor. Other voices added to the national despair. Bernard M. Baruch, a venerated public intellectual, called Americans to a new regimen of hard work, to abandon the passion for Detroit's automobiles, "the gaudy, grinning, chrome-plated, tail-finned, wrap-around, Dynaflowing embodiment of . . . moral and spiritual flatulence." Just hours after Sputnik 1 reached orbit, Margaret Mead, the famed American anthropologist, created her own "emergency Sputnik survey," designed to collect raw data on the popular reactions to the Soviet space triumph. Joined by Rhoda Metraux, Mead would seek out the reactions to Sputnik from more than 5,000 individuals throughout the United States, Canada, and Hawaii.[19]

Homer H. Hickam, author of the best-selling book *Rocket Boys,* mirrored as a 14-year-old the powerful passions ignited by Sputnik as a talisman of the space age. Hickam, who later pursued a career as a NASA engineer, observed Sputnik from the vantage point of his home in West Virginia. From his backyard, he became enthralled with the "bright little ball, moving majestically across the narrow star field between the ridgelines." Awestruck by the satellite, a human creation catapulted into the heavens, Hickam noted that he stared at the orbiting ball "with no less rapt attention than if it had been God Himself in a golden chariot riding overhead. It soared with what seemed to me inexorable and dangerous purpose, as if there was no power in the universe that could stop it."[20] The young Hickam participated with countless millions who gathered on rooftops, in parks, and in backyards to catch a glimpse of the technical marvel of the age.

Another American moved by the Sputnik event was Neil Armstrong, the future spacecraft commander of Apollo 11 and the first human to step on the surface of the moon. On that fateful day of October 4, Armstrong was in Los Angeles, California, attending a meeting of the Society of Experimental Test Pilots. As a member of this small fraternity, he had long lamented the general indifference of the media toward test pilots, even those involved with upper atmospheric flying with the X-15 rocket plane. For Armstrong, "what was happening in the test-flight world was a very hard sell to the press, and it became completely impossible once Sputnik came across the sky." Later, he reflected on the larger meaning of October 4: "Sputnik did change our world. It absolutely changed our country's view of what was happening, the potential of space. I am not sure how many people realized at that point just where this would lead. President Eisenhower was saying something like, 'What's the worry? It's just one small ball.' But I'm sure that was a facade behind which he had substantial concerns, because if they could put something into orbit, they could put a nuclear weapon on a target in the United States." By 1962, Armstrong had abandoned the test-flight community for training as an astronaut: "I decided that if I wanted to get out of the atmospheric fringes and into deep space work, that was the way to go." Sputnik had redefined the future—rockets, not winged craft, were now at the cutting edge of experimental flight.[21]

For the editors of *The Saturday Review,* the annoying beeps of the Sputnik satellite indeed had mocked America's "delusions of superiority."[22] However, this setback did not necessarily mean that there was some inadequacy in American science and technology. According to this analysis, America should undertake no crash program in arms development or redesign the existing educational curriculum of schools and colleges. One such voice was the poet Archibald MacLeish, former head of the Library of Congress. Appearing on the campus of the University of Kansas in the immediate aftermath of Sputnik, he dismissed the whole idea of ballistic missiles as an "idiotic dream." Moreover, MacLeish reacted negatively to any "crash program to keep up with Russia," arguing that "we are going to end up with a lot of plumbers and electrical helpers." He expressed profound fears about the growing clamor for more technological training, a false remedy for what he viewed as an illusory problem—a notion that would exalt mastery of technique over all else, creating a new generation of "drones in the beehive . . . [who] will get stung in the end."[23]

Two weeks after the launch of Sputnik 1, poet and historian Carl Sandburg appeared on NBC's *Meet the Press,* where the questions turned from pressing domestic issues to the matter of the Soviets launching the first Earth satellite. White-haired, then 79 years old, and widely revered as the eminent biographer of Abraham Lincoln, Sandburg took the occasion to renew his criticism of the Eisenhower administration. The Sputnik crisis, he pointed out, revealed certain tendencies in modern American life that had allowed the Russians to forge ahead in the sphere of space, such as "McCarthyism, philistinism, and neglect or lack of respect for sciences."[24]

Some prominent political figures, however, saw the crisis in more conventional terms, arguing that science and technology had faltered in the United States. Former President Harry Truman took the occasion of a Jefferson-Jackson Day dinner to join the chorus of strident critics of the Eisenhower administration. Speaking before an enthusiastic audience in Los Angeles on November 1, 1957, Truman noted that the Russians had "scored their greatest propaganda victory in many years." The Sputnik achievement had taken place "while our Secretary of State [John Foster Dulles] dances on the brink of war, and the President cuts down the Army, the Navy, and Air Force." In Truman's assessment, President Eisenhower had been "slow to sense the new Russian danger, and when made aware of it, he was slow to do what he ought to do." After these general partisan utterances, Truman identified the true culprit: "While the Russians were demonstrating their scientific advances, the administration was actually issuing secret orders cutting down the basic research in the defense establishment . . . serious blunders, shocking in their lack of judgment, imperiling the country's safety."[25]

Amid the intense debate on the Sputnik launches, a clamor rose for the United States to launch its own Earth satellite. This demand became loud and demanding, forcing the government to announce that the first Vanguard launch would take place on December 6, 1957, as part of America's contribution to the IGY celebration. This decision reflected the pressures of the moment, since only the first of the four stages of the Vanguard rocket had been tested successfully.

Given the fact that two Soviet satellites were in orbit, the Vanguard launch attracted a global audience. At the moment of ignition, Vanguard's engines roared to life, thrusting fire and smoke downward, giving a momentary feeling that the black-tipped rocket would lift majestically toward the heavens. Those hopes were soon crushed as the Vanguard—after rising a mere four feet—collapsed onto the launch pad in a cloud of fire and smoke. Millions watched the unfolding debacle in amazement. In the debris field of burning fuel and wreckage, the Vanguard satellite capsule somehow managed to survive intact, sending its signals from the charred ground zero. This embarrassing moment prompted the writer Dorothy Kilgallen to say, "Why doesn't someone go out there and kill it?" Tom Wolfe, the author of *The Right Stuff*, described the scene at Cape Canaveral in devastating detail: "The first stage, bloated with fuel, explodes, and the rest of the rocket sinks into the sand beside the launch platform . . . very slowly, like a fat man collapsing into a Barcalounger. . . . This picture—the big buildup, the dramatic countdown, followed by the exploding cigar—was unforgettable."[26]

The foreign press, to the chagrin of the Eisenhower administration, was equally dismissive of Vanguard, regarding the calamity as a testament to the moribund state of America's rocket program. No less troublesome was the shorthand widely used to describe the ill-fated Vanguard—words such as "Kaputnik" or "Stayputnik." The rocket project itself was dubbed derisively as "Rearguard." The United States greeted the arrival of 1958 in a mood of national depression. It was tempered, however, by a growing desire to catch up with the Russians.

NARROWING THE GAP

As unlikely as the coincidence might seem, on the night Sputnik 1 was launched, von Braun and General Medaris were entertaining Neil McElroy, the defense secretary-designate, and Army Secretary Wilbur Bruckner at the Redstone Arsenal. As the news of Sputnik 1 came in, von Braun wasted no time in making clear to these powerful visitors his disdain for Vanguard and its prospects for success and his frustration that "[w]e could have done this with our Redstone two years ago. . . . We knew they [the Soviet Union] were going to do it." Well aware of Vanguard's problems and delays, he said, "Vanguard will never make it.

We have the hardware on the shelf. For God's sake, turn us loose and let us do something." Von Braun did not stop there: "We can put up a satellite in 60 days, Mr. McElroy." He repeated "60 days" several times before being interrupted by Medaris: "No, Wernher, *90* days." McElroy made no commitment before returning to Washington.[27]

But Medaris and von Braun did not have long to wait. On November 8, 1957, just five days after the Soviet Union's second space spectacular—the launch of Sputnik 2, carrying the first living creature into orbit—the U.S. Army finally received approval to launch a satellite. In making his decision, Defense Secretary McElroy, who had himself just taken office days earlier, recalled the confident can-do attitude voiced by Medaris and von Braun in Huntsville. He informed Medaris, who quickly gave von Braun the news: "Wernher, let's go!"[28]

Beyond readying one of its Jupiter-C missiles for launch, a major challenge for ABMA was the prompt development of the satellite it would carry. The Army team wanted a scientific payload that was considerably more sophisticated than that envisioned for Project Orbiter. Jet Propulsion Laboratory (JPL) director William Pickering convinced General Medaris that his institution, which had been involved in Project Orbiter from the beginning, should retain its role going forward. JPL was awarded the contract for Explorer, as the new satellite was soon named. The satellite was an 80-inch-long, six-inch-wide cylinder, unlike Sputnik 1 and Vanguard, which were spheres. Explorer, weighing just over 30 pounds, was divided into two sections: one containing measuring instruments and radio transmitters totaling 18.5 pounds and the other section a solid-propellant rocket to kick it into orbit.[29]

Of the three experiments onboard, by far the most significant was the one contributed by James Van Allen, in whose Maryland home the IGY had been conceived. Soon after that meeting, Van Allen had moved to the University of Iowa, where he headed the department of physics and astronomy. His Explorer experiment was a special Geiger counter that would bring him international fame by discovering two separate radiation belts of charged particles (cosmic rays) trapped by Earth's magnetic field between 400 and 15,000 miles above the planet. The belts, named for Van Allen, constituted the first significant scientific discovery of the space age, and heralded the coming age of human exploration of the solar system.[30]

In 1958 the United States prepared to establish itself as a serious rival to the Soviet Union. With the launch of Explorer 1, it hoped to achieve its own milestone by firing an artificial satellite into orbit. For the American scientific community, Explorer 1 would become an important tool for scientific research.

Von Braun and his team were more than ready for their hard-earned shot at orbiting the first American satellite. They had also learned some lessons from the Navy's disaster. In the first place, Medaris ordered the Army launch plan to operate under tight security. For example, the launch rocket was referred to only as "Missile 29" in classified Army communications, to make it appear it if it were to be used for just another advanced Redstone missile test. The missile was flown to Cape Canaveral in late December 1957, its upper stages discreetly covered in canvas to hide their shape. The launch missile's name was changed as well, but that wasn't the result of Medaris's security regime. Rather, in deference to Eisenhower's concerns about the use of military missiles for IGY activities, the Jupiter-C's used to launch satellites were known as the Juno 1.[31]

All was ready on the nights of January 29-30, 1958, but the weather wouldn't cooperate, with strong winds at higher altitudes ruling out a launch. Conditions were better the next night, and just before 11 P.M. the Juno 1 was headed for space. General Medaris was at the Cape to view the launch, but he was hardly alone. Despite the tight security he had ordered, word had gotten out, and thousands of cheering onlookers enjoyed the view from nearby beaches. Von Braun waited in the Pentagon's communications center, with Pickering and Van Allen.[32]

The Juno consisted of four stages, including the Jupiter-C first stage. Fourteen small Sergeant rockets made up the remaining stages, arranged in a drum-like container, with 11 in the outer ring (second stage) and 3 inside the ring (third stage), with Explorer on top of them. Firing in the proper sequence, the first three stages would exhaust their fuel and then drop away after burn-out. Finally, the fourth stage, Explorer itself, would be boosted into orbit by the final Sergeant. After the rocket blasted skyward that night, there was a longer-than-anticipated interval in receiving final confirmation that Explorer 1 had achieved orbit. Attributed to various causes, such as the satellite being temporarily out of radio range of the nearest tracking station, the wait was finally rewarded with positive news: Explorer 1 was a success![33]

"None of the [tracking] stations were hearing a thing," von Braun later recalled. "That went on for what appeared to be hours. [The time was actually much less than that.] Meanwhile, we had to keep up appearances and had to smile and convince everybody that things were in perfect shape." Eisenhower made a brief public announcement that "[t]he United States has placed a scientific satellite in orbit around the Earth" as part of its IGY participation. Privately, he told the aide who first brought him the news of Explorer's success, "That's wonderful. I sure feel a lot better now."[34]

Explorer 1 flew in an elliptical orbit, varying from 220 miles at its closest point to Earth to as far as 1,563 miles above the planet. It orbited Earth every 114.8 minutes, or a total of 12.54 orbits daily, and achieved more than 58,000 orbits before reentering Earth's atmosphere and burning up on March 31, 1970. It continued to transmit data from orbit until May 23, 1958. Eisenhower and the entire nation felt the immense satisfaction that the American space program had achieved a real measure of success. In March 1958, Vanguard would finally launch its own orbital satellite, followed by the second successful Explorer, Explorer 3, on March 26. (Earlier in March, Explorer 2 did not achieve orbit due to a failure in the fourth stage of its Juno 1 launch vehicle.)

The successful orbital launch led to the fulfillment of another dream as well. Once "freedom of space" had been established by Sputnik 1's orbits across the United States, and America lofted its own IGY scientific satellite, Explorer 1, Eisenhower moved to exploit space reconnaissance in February 1958. He approved an intermediate spy satellite system to serve U.S. requirements pending full development of the WS-117L system he had authorized earlier. The new system, named Project Corona by CIA officials, involved taking pictures from space and then returning the film to Earth using a capsule detached from the satellite. It turned out to be a vital and highly successful element of American efforts to learn what the Soviets were up to on their side of the Iron Curtain.

For both the Americans and Soviets, the space race was now in full stride.[35]

REACHING FOR THE MOON

Spurred on by the American successes, Korolev was anxious to catapult the Soviet Union into a new realm of space exploration—a bold plan to abandon the near-Earth orbits of the Sputnik satellites for a series of lunar probes. He prepared a paper on lunar exploration, a detailed blueprint of the essential steps required to reach the moon. His elaborate scheme took into account all the anticipated technical challenges: the rocket design, payload limits, necessary instrumentation, and launch windows. He put forth a compelling rationale: The deep space robotic missions offered yet another, even more dramatic, avenue to display the country's prowess in space exploration. No less important, the projected lunar probes served an important scientific purpose—the unique opportunity to measure a range of outer space mysteries, from the moon's magnetic field to cosmic radiation to micrometeorites. The most formidable challenge, as with the earlier Sputnik launches, was the design of a rocket with sufficient thrust to escape Earth's gravity. The R-7 stood ready, but the veteran rocket would have to be fitted with an additional third stage to achieve the necessary speed of approximately 7 miles per second. No less daunting would be the design of an advanced guidance system to ensure that the rocket maintained a precise course in the void of outer space.

Standing at the top of the ziggurat of Soviet politics was Nikita Khrushchev, and his posture toward these civilian uses of rocketry became the critical factor. He had approved the deployment of the R-7 for the Sputnik 1 and Sputnik 2 with great enthusiasm. Khrushchev displayed an equal measure of openness aboout the lunar probes, seeing the enormous propaganda potential in Korolev's bold plan. Khrushchev—in the words of his son, Sergei—"wanted to beat the Americans in all spheres of life and to prove that our socialist system is working better." Space was just one component of this overarching strategy.[36]

His devotion to rocketry grew with each successive triumph. As premier of the Soviet Union, Khrushchev took delight in the public spectacle of space milestones. He had frequent conversations with designers and, later, with the growing cadre of cosmonauts as the Soviet manned space program took shape in the 1960s. His tenure in power endured until late 1964, when he was ousted in a coup led by Leonid Brezhnev.

Korolev's lunar probes, once approved in March 1958, started as a frustrating blend of delays and abortive launches in the early months of the program. Engine malfunctions were accompanied by acrimonious debates about the optimal engine configuration between Korolev and Valentin Glushko, the premier Soviet rocket engine designer. On September 23, 1958, there was an abortive launch of a Luna spacecraft, when the booster disintegrated just seconds into the liftoff. This same melancholy script was repeated on October 11, just days after the first anniversary of Sputnik 1. Another embarrassing failure followed on December 4, which only deepened the mood of futility at Baikonur.[37]

Success finally came with Luna 1 (named *Mechta*, "dream"), launched on January 2, 1959. The redesigned R-7 successfully lifted the lunar probe beyond the grasp of Earth's gravitational pull and set it on a fateful course toward the moon. After reaching escape velocity, the probe separated from the third stage. At the 70,000-mile mark, the spacecraft released a two-pound cloud of sodium gas, creating an extended orange vapor trail. This streaming patch of light with the brilliance of a sixth-magnitude star was visible over the Indian Ocean, allowing for a visual confirmation of the lunar trajectory.

Luna 1 was designed to be the first human-made object to reach the surface of the moon, but the historic collision with the celestial body did not occur: The spacecraft passed the moon 34 hours after launch, missing its target by some 3,700 miles. Luna 1 then went into orbit around the sun, trumpeted as another first for the muscular Soviet space program. While Luna 1 did not carry any cameras on board, it had been fitted with sophisticated instruments, including detectors that measured Earth's radiation belt, which demonstrated that the moon had no magnetic field and discovered solar wind, the powerful flow of ionized plasma emanating from the Sun. Luna 1 represented a genuine milestone, one that moved the arena of competition between the Soviet Union and the United States into deep space.

The Soviets launched Luna 2 on September 12, 1959; it became the first Soviet probe to hit the moon. As with the first lunar probe, a large orange cloud of sodium gas was released, not just to mark the spacecraft's trajectory, but to study the gas in the weightless environment of outer space. The lunar probe Luna 2, with a payload of 800 pounds, reached the moon after nearly a day and a half in flight, crashing into the lunar surface at a point east of the Sea of Serenity. Upon impact, the probe scattered a number of metal pendants to mark the spot of the historic impact. The pendants were inscribed with a simple message: "U.S.S.R. September 1959." Hitting the moon—admittedly a huge but a fast-moving object approximately 240,000 miles beyond Earth—represented for the time an impressive display of marksmanship. The Soviet lunar rocket embodied a sophisticated guidance and navigation system to achieve this end.

The most impressive lunar probe, Luna 3, departed Earth on October 4, 1959, on the second anniversary of Sputnik 1. This launch represented the most ambitious maneuver into outer space yet attempted by Korolev and his team of engineers and scientists. Taking leave of Baikonur, Luna 3 entered a highly elliptical orbit from Earth, in a figure-eight trajectory that took it around the moon and then back again toward Earth. Coming as close as 3,800 miles to the moon's surface, Luna 3 sped around to the far side (always hidden from view on Earth) to take photographs. On October 7, the television system on the probe took 29 photographs covering 70 percent of the moon's far side. The photography took 40 minutes to complete. The images were then scanned for transmission to ground stations on October 18, at an optimal moment when Luna 3 neared Earth on its return path. The Soviets collaborated with the British to allow the famed 250-foot radio telescope at Jodrell Bank to record the pictures sent back from the Luna 3 spacecraft, a remarkable request given the secrecy surrounding all launches from Baikonur. The processed images—the first ever taken of the "dark side" of the moon—were not high quality, but they allowed for mapping and scientific analysis of the hitherto hidden features of the lunar surface.

Having captured these unique images, the Soviets promptly christened certain prominent features of the newly discovered lunarscape, for example two "maria" (Latin for "seas") on the far side of the moon were dubbed the "Sea of Moscow" and the "Sea of Dreams." Throughout the dramatic lunar missions, Korolev had once again displayed his skill and efficiency in pushing the Soviet space program toward epochal new milestones. There was great irony in this latest chapter of Korolev's career, the fact that he had enabled the Soviet Union to gain worldwide prominence as a pioneer in space exploration, yet he remained an anonymous figure in Soviet official media.[38]

After the successful launch of Explorer 1 in 1958, the United States also turned its attention to lunar probes, seeking to compete in this new realm of space exploration. Initially, the primary aim was to design rockets to make flybys or possibly to impact on the lunar surface. An additional key factor in the development of lunar probes would be experimentation with new imaging systems.

Pioneer 1, launched on October 11, 1958, ended in disaster, as the rocket failed to achieve escape velocity. Pioneer 2 followed on November 8, a mission that ended when the rocket's third stage malfunctioned after an ascent of less than a thousand miles. On December 6 of that same year Pioneer 3, powered by a Jupiter-C Juno 2 rocket, traveled about one-fourth the way to the moon. However, Pioneer 3's trek into outer space ended abruptly after it suffered a premature engine shutdown; the rocket ignominiously fell back into Earth's atmosphere. Yet the saga of Pioneer 3 was not without scientific meaning: The abortive mission, before its fiery demise, confirmed the existence of the Van Allen Belts.

These formative steps by the United States were quickly overshadowed in early January 1959, when Luna 1 executed the first-ever lunar flyby. In response, the decision was made to launch Pioneer 4 on March 3. Finally, American persistence was awarded with a measure of success; Pioneer 4 executed a flyby of the moon. Although a first for America's troubled space program, the mission evoked an image of misadventure: The rocket veered off its planned path by 37,500 miles—an unscripted trajectory caused by an engine malfunction.

At this juncture, the Soviet payloads for lunar probes outclassed the Americans in a dramatic fashion: while Luna 1 weighed around 796 pounds, Pioneer 4 tipped the scales at a puny 13 pounds! The American lunar probe missed the moon by a wide angle and then entered into a solar orbit. In 1960, Pioneer 5 carried out a successful interplanetary mission, sending its final radio message back to Earth across a distance of 22 million miles. From November 1959 to December 1960, the United States launched three lunar probes on Atlas Able boosters. All failed in their mission to reach lunar orbit.[39]

Even with these reversals, the American space program was slowly acquiring maturity, building the critical institutional and technical means for future success in robotic missions to the moon and beyond. One vital research center for this activity in the 1960s would be the Jet Propulsion Laboratory (JPL) in Pasadena, California. The laboratory dated back to the late 1930s, being an outgrowth of Theodore von Kármán's research at the California Institute of Technology. Along with von Kármán, Frank J. Malina worked to make the JPL a major center for research.

William Pickering emerged as a leader of JPL in the 1950s, giving the lab increased stature as a center for research and development of satellites, robotics, and advanced guidance systems. As it had developed Explorer 1, leading America into space, JPL would take the lead in the Ranger program in the 1960s, an important cornerstone of America's Apollo manned lunar landing program. After several failures, the Ranger program finally yielded a success with Ranger 7 in July 1964. The probe returned 4,000 stunning photographs of the moon's surface before crashing into it, helping to pave the way for manned landings. The gap with the Russians in all spheres of space travel would be closed as the decade of the 1960s ended.

NASA TAKES CHARGE

The opening salvoes of the space age brought the realization that some new organizational form or structure would be required to respond to the Soviet space challenge. Space was a new and different realm, a fact understood by all of the major players, including the Eisenhower administration, Congress, the military, defense contractors, and other involved parties. Eisenhower's response was to propose to Congress the creation of a new federal agency, the National Aeronautics and Space Administration (NASA). He proposed that the new entity absorb a small existing agency, the National Advisory Committee for Aeronautics (NACA) and its various facilities. NACA, created in 1915, had evolved from conducting only aeronautical research to contributing significant space-related work by the opening of the space age. However, it was regarded as too small an agency to take on the additional responsibilities envisioned by Eisenhower and his science advisor, James Killian.[40]

Eisenhower and Killian strongly believed that the national space agency should be civilian, but the Army and the Air Force resisted fiercely, asserting that they should be in charge of the space program, and that turning it over to a civilian agency would be a mistake. General Medaris and von Braun, among others, were strong advocates of this position. The Air Force was just as aggressive, and one of its generals publicly advocated a military base on the moon in an early 1958 speech. "The moon provides a retaliation [sic] base of unequaled advantage," he said. A nuclear-armed American base on the moon would allow the United States to inflict "from the moon . . . sure and massive destruction" on the Soviet Union in the event it first attacked the United States, he added.[41]

Eisenhower's proposed legislation creating NASA was subjected to intense congressional debate, ending with the agreement with the president's view that a civilian agency would oversee scientific and other nonmilitary space efforts, while reserving military requirements and programs for the armed services. But enough ambiguity remained to keep the space missions issue alive for years to come in such areas as control over huge and powerful military rockets that would be required to support civilian programs.[42] In the short run, though, the legislation creating NASA, submitted to Congress by the White House in April 1958, was signed into law by Eisenhower in July of that year.

A second noteworthy law took shape following a major debate over whether America was producing enough engineers and scientists to meet the Soviet challenge in space and elsewhere. The debate started immediately following the surprise of Sputnik 1 and focused on what role the federal government should play in closing the perceived educational gap between the United States and the Soviet Union. The National Defense Education Act, signed by President Eisenhower in September 1958, was intended as a limited and temporary measure. It made generous grants for the expansion of programs aimed at students studying science, mathematics, engineering, and foreign languages in high schools, colleges, and universities.[43]

The United States was now truly in the space game. It had recovered from the shock of Sputnik 1 with a satellite of its own and more, and seemed ready to compete with the Soviets in the new sphere of unmanned lunar probes. But these were only the earliest rounds of space exploration. Many challenges and surprises were ahead for both sides as they moved into the 1960s and the era of manned spaceflight, first in Earth orbit and then far beyond.

EYES IN THE SKIES

In early 1955, more than two years before Sputnik 1 inaugurated the space age, a high-level panel delivered a top-secret report to President Eisenhower. It strongly recommended that the United States use Earth satellites to provide accurate intelligence from space on the true state of Soviet offensive capabilities. As a result, the CIA and the U.S. Air Force soon began to develop photo-reconnaissance satellites under a highly classified program code-named Corona.

Corona operated under the cover name "Discoverer" and the accompanying cover mission of scientific research. Following a series of heartbreaking failures, Corona finally delivered the goods: high-quality photographs taken from space of selected sites in the Soviet Union. Discoverer satellites were launched from Vandenberg Air Force Base in California, allowing them to be placed in polar (north to south) orbits, during which the entire globe

An Air Force C-119 recovers, in mid-air, spy photographs taken over the Soviet Union.

passed below them. The Discoverer's camera was turned on as the satellite passed over pre-selected reconnaissance targets in Russia or elsewhere. Assuming that all went well to that point, the remaining challenge was to return the photos, taken on strips of 70-mm film, successfully to Earth. That was accomplished in a unique manner: A reentry capsule containing the exposed photographic film was detached from the remainder of the Discoverer satellite and rocketed back toward Earth.

After successful reentry, a parachute opened and one of a group of Air Force planes, each deploying a flying circus-like trapeze, or net, was to ensnare the parachute in midair and then reel it into the aircraft. Not surprisingly, given the extraordinarily complex nature of all aspects of the Discoverer program, success was elusive for the first dozen plus attempts.[44]

Finally, in August 1960, Discoverer 14 ended its mission successfully, completing a unique high-wire act as its reentry capsule was captured in a perfect mid-air recovery. The capsule's film yielded photographs fit to show a president, and that is precisely what CIA Director Allen Dulles did personally in the Oval Office on August 24. A reel of developed Discoverer film was unrolled at Eisenhower's feet. A sampling of the satellite's total mission "take," it showed high-quality photographs of Soviet and Eastern European territory taken from 115 miles above Earth.

Overall, the photos showed dozens of air bases and surface-to-air missile sites and a major new Soviet rocket facility at Plesetsk, in northern Russia, where a very small number of Soviet R-7 ICBMs were later based. In sum, Discoverer 14 had provided more useful intelligence information than four years of U-2 flights (which had ended in May 1960 after a U-2 was shot down over Russia). The photographs showed objects as small as 6 to 9 feet.

The significance of all this was summed up in a later comment by Albert "Bud" Wheelon, the CIA's deputy director for science and technology: "It was as if an enormous floodlight had been turned on in a darkened warehouse."[45] Other Discoverer satellites continued to bring back photographic proof that, despite Khrushchev's boasts that Russia was producing ICBMs "like sausages," there were *no* hidden caches containing hundreds of Soviet ICBMs ready to strike the United States without warning. ■

THE HUMAN DIMENSION

A manned space mission represented the next logical step in the American space program—what became known as "Project Mercury." T. Keith Glennan, administrator of the newly organized National Aeronautics and Space Administration (NASA), chose a day filled with symbolic meaning to unveil this new initiative—December 17, 1958, the 55th anniversary of the historic flight of the Wright brothers at Kitty Hawk. For NASA planners, these future space travelers, soon to be known popularly as "astronauts," would ride into outer space in a specially designed spacecraft and then return safely to Earth. For the first time, there would be a human presence in outer space.

There was also a thinly veiled subtext: The Mercury program reflected the determination of NASA to place a human in orbit—and ultimately the pursuit of a lunar landing—ahead of the Soviet Union. Most Americans believed that the Soviets possessed hidden capabilities to achieve the same lofty goal, given their recent space exploits. By contrast, the American space program appeared backward, seemingly adrift and without focus.

OPPOSITE: *Soviet cosmonaut Yuri Gagarin sits at the ready at Baikonur cosmodrome shortly before his launch into orbit, April 1961.*

There was a new dynamic at play, one calling for a race with the Soviets for prestige and preeminence in space. Given this emerging space rivalry with the Soviets, President Eisenhower—to the frustration of the NASA leadership—persisted in a go-slow approach. He had warmly endorsed the goal of launching artificial satellites, which was justified in his mind as a genuine scientific undertaking and in harmony with his own goal of affirming the freedom of space. Moreover, for Eisenhower, satellites offered the United States an effective platform for aerial reconnaissance—in the context of the Cold War, a vital function to assure America's national security. Consequently, Eisenhower gave human spaceflight a lower priority, even though he had allowed Project Mercury to take shape. In his last budget to Congress, he recommended that no human spaceflights be pursued beyond Project Mercury unless they possessed "valid scientific reasons."[1]

Eisenhower had been instrumental in forging America's missile technology and even laying down the institutional foundation for the American space program. His interest in space, however, had remained narrow, for the most part tied to the need for reconnaissance satellites. He had no appetite for space spectaculars. In the immediate aftermath of the Sputnik launches, a chorus of critics arose to accuse the Eisenhower administration of inertia and drift. Efforts by Eisenhower to reassure the American public proved largely ineffectual. Being privy to highly classified intelligence reports, Eisenhower was aware of the technological weaknesses of the Soviet Union—and, ironically, the advanced nature of American Atlas, Titan, Polaris and Minuteman missile programs. He fully understood that there was no missile gap. Moreover, he rejected the idea that the Sputnik satellites mirrored some sort of strategic advantage for the Soviet Union. He feared that any embrace of a long-term "space race" with the Soviets would be foolhardy and unnecessary.

The Eisenhower approach reflected a strong preference for robotic probes into outer space for scientific research. By contrast, the pursuit of manned programs—into orbit and beyond—required greater expenditures and prodigious engineering feats. For Eisenhower, any ambitious program for the human exploration of space would be costly,

Sergei Korolev (right) sits in a glider designed by B. I. Cheranovsky (left), circa 1930s.

1945 London V-2 attack inflicted hundreds of casualties.

OPPOSITE: *German slave laborers build Nazi V-2s in 1944.*

NEXT PAGE: *Test launch of a German V-2 in White Sands, New Mexico, in 1951.*

Soviet engineers, including Sergei Korolev, with first liquid-fueled rocket, 1933.

Wernher von Braun and colleagues after the 1945 surrender to the U.S. Army in Germany.

Laika's 1957 ride on Sputnik 2 made her the first living creature to orbit Earth.

OPPOSITE: *Chimpanzee Ham's 1961 suborbital space flight helped pave the way for humans.*

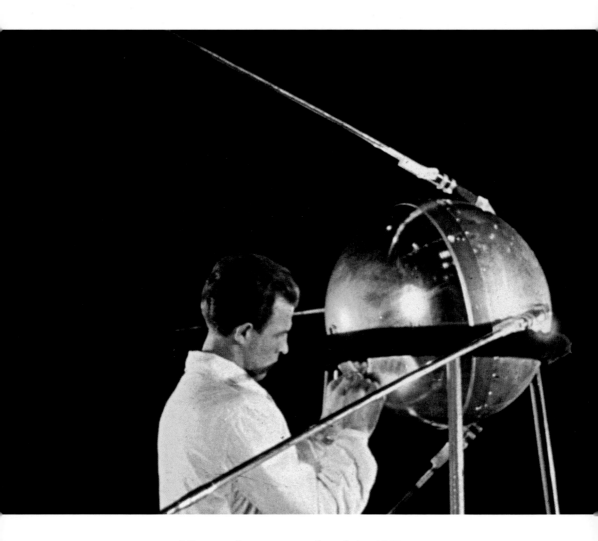

A Russian technician prepares Sputnik 1 in 1957.

Opening a new era: the launch of Sputnik 1, October 4, 1957.

NEXT PAGE: *Thrill of the new: Chicagoans seek Sputnik 1 sighting.*

TOP and BOTTOM: *Russia builds a new spaceport: Baikonur, mid-1950s.*

Vanguard, December 1957: a disastrous end for America's first orbital attempt.

Children of the new age: firing toy rockets, late 1950s.

requiring a vast expansion of the aerospace industry. This position became increasingly untenable in the post-Sputnik environment. The voices clamoring for the United States to aggressively challenge the apparent Soviet superiority in space only grew in intensity as the decade ended. Project Mercury, if defined narrowly under Eisenhower as a program for manned spaceflights in near-Earth orbits, would become the seedbed for a more ambitious American endeavor in space, one that would rival in its scope the Manhattan Project of World War II. The key force behind the shift in national policy would be John F. Kennedy, who would be elected to the presidency in 1960.

NASA became proactive in promoting the American space program, even with only a tepid display of support from the Eisenhower White House. At the core of the new NASA organization was the old National Advisory Committee for Aeronautics (NACA), a venerable and highly productive research entity that traced its origins back to 1915. Building on these NACA roots, NASA took under its wing a number of key research organizations, including the Langley flight research center (soon to become the home for Project Mercury); the Lewis Flight Propulsion Laboratory in Cleveland, Ohio; the high-speed flight center at Edwards Air Force Base; and the Wallops Island test facility in Virginia, among others. In October 1958, the Jet Propulsion Laboratory (JPL) came under NASA, as did much of the army rocket program at Huntsville in the fall of 1960, eventually becoming the George C. Marshall Space Flight Center. With the latter transfer, Wernher von Braun found himself under a civilian agency with a clear mandate for space exploration.

During the years 1958-1960, NASA attempted a number of satellite launches. Success was eventually attained, but only after a number of embarrassing mishaps. The four attempts in 1958 all ended in failure. Another 14 launches in 1959 garnered nine successes. The year 1960 showed greater progress, with a record of 12 successful tries in 17 attempts. During this same period the Soviets scored a series of remarkable successes with their Luna series: Luna 1, the first to range outside Earth's gravity; Luna 2 crashing into the Moon's surface; and, most memorable of all, Luna 3 taking photographs of the far side of the moon. Still, NASA could point with pride to launching the first weather satellite, TIROS 1, in April 1960, and the first navigation satellite, Transit 1B, the same month, among other benchmarks.[2]

NASA administrator Glennan, the past president of Case Institute of Technology and a prominent member of the Atomic Energy Commission, proved to be an effective administrator in consolidating a number of diverse research entities under the NASA umbrella. But others also played key roles in shaping the emerging American space policy.

Hugh L. Dryden, the long-term NACA leader, had displayed a keen interest in promoting a viable space program in the United States. He had led the high-flying X-15 test program in the late 1950s. He also played a pivotal role in mapping the future space program while still at NACA. Dryden appointed Robert Gilruth, the hard-driving head of the Langley research facility, to assemble a panel (to be known as the Space Task Group) to formulate a long-range program for spaceflight. The work of Gilruth and his special panel set out the essential architecture for Project Mercury. Both Dryden and Gilruth would join Glennan at NASA headquarters.[3]

For Project Mercury, there was a compelling need to fashion a whole new cluster of technologies to allow for human spaceflight. Any vehicle carrying humans into space had to be capable of resisting the perils of outer space—extremes in temperature, the airless void of space, and the newly discovered dangers of radiation. The optimal design for a capsule prompted debate and competing designs. Some argued for the older configuration of an aircraft, to fly into space and then glide back to Earth. The X-15 provided one such model, since it could reach an altitude of 67 miles, but it lacked the necessary acceleration to enter space on its own power. H. Julian Allen, who had established himself as a leading NACA engineer in the 1950s, argued for a missile to launch a blunt-shape capsule, one that would resist the heat generated upon reentry and still provide the requisite aerodynamic properties to assure a safe landing.[4]

However, the preferred option, approved by Glennan, originated with Maxime Faget, an engineer at Langley. He argued for a capsule with retro-rockets to slow the orbital momentum, allowing for a gradual descent into the atmosphere. The final design reflected this basic approach—a cone-shaped capsule with a cylinder at the top (then fitted with a special tower containing a solid-propellant rocket to fire and lift the capsule to safety in case of any emergency on the launch pad). The capsule was rather small, offering its occupant little comfort while in orbit, and the astronaut sat in a custom-designed seat. Eighteen small rockets, which were operated manually by the astronaut, controlled the attitude of the capsule and directed the spacecraft into a home trajectory from orbit. The capsule's ablative heat shield protected the astronaut during reentry into the atmosphere. Parachutes slowed the capsule during final descent for a splashdown in the ocean. It is interesting that these far-reaching plans were articulated before July 29, 1958, when President Eisenhower signed the legislative act to create NASA. Finally, McDonnell won the contract for the development of the capsule on January 12, 1959.[5]

NASA employed different booster rockets to catapult the Mercury capsule into space. The "Little Joe" rocket served as a booster to launch dummy capsules into the upper atmosphere for testing. The venerable Redstone, in reality an upgraded variant of the V-2, was deemed ideal for sub-orbital flights of dummy capsules and eventually astronauts. For the final step, the launching of an astronaut into an Earth orbit, NASA decided to use the Air Force's powerful Atlas rocket. The Atlas, still in development in 1958, would be fitted with engines capable of 180 tons of thrust. The Atlas—a massive ICBM with a range of 9,000 miles—offered the astronaut an extraordinary ride into space at 17,500 miles per hour. General Dynamics had been given the primary contract for the development of the Atlas.[6]

In the second half of 1958, NASA began recruiting the first class of astronauts. Gilruth's Space Task Group assumed the leadership for this selection process. The initial job description, written up in the framework of a civil service position, cast a rather wide net for hard-nosed adventurers—one could qualify from myriad pursuits such as submariners, parachute jumpers, and Arctic explorers. This open-ended appeal soon was abandoned in favor of a more narrow set of criteria, insisted upon by President Eisenhower in late December 1958: The astronaut corps had to be recruited from the community of military test pilots on active duty. Now refocused, the selection process proceeded forward. Only males between the ages of 25 and 40 could apply. They could be no taller than 5 feet 11 inches, and had to possess a college degree or equivalent experience in an appropriate technical field. Flying experience became an essential factor, a minimum of 1,500 hours with extensive cockpit time in jets. The initial pool consisted of some 110 candidates, which was distilled down to 69 applicants who weathered an intense round of tests and interviews. The relentless process of whittling down continued until the group was reduced to seven.[7]

The selection of these seven men, the nation's first class of astronauts, was announced to the public on April 9, 1959. Reflecting Eisenhower's narrow criteria, they all came from the ranks of the military, all were aviators, some with considerable experience as test pilots: Marine pilot John H. Glenn; Navy aviators Alan B. Shepard, M. Scott Carpenter, and Walter M. Schirra; and Air Force fliers Virgil "Gus" Grissom, L. Gordon Cooper, and Donald "Deke" K. Slayton. Their debut, to NASA's surprise, created a sensation. The public expressed a keen interest in the new corps of astronauts, seeking to know more about their lives, families, and interests. The Mercury Seven became immediate celebrities, the object of an adoring populace and media. A year after their debut, the astronauts signed a contract giving *Life* magazine the exclusive right to tell their personal stories. The public image of the Mercury Seven reflected a blend of patriotism, bravery, and technical competence. They were the heroes of the new space age—at the cutting edge of America's campaign to gain dominance in the space race with the Soviets. Novelist Tom Wolfe, in his memorable bestseller *The Right Stuff*, captured this moment: ". . . all seven, collectively, emerged in a golden haze as the seven finest pilots and bravest men in the United States. A blazing aura was upon them."[8]

Less apparent to the public and the adoring media was how problematical "flying" in the Mercury program became, at least initially, for these seasoned pilots. Seated and wired, the astronaut was catapulted into space in a highly automated flying machine. He had minimal opportunity to exercise manual control in the spacecraft. The elite fraternity of test pilots, most notably fabled airmen Chuck Yeager and Scott Crossfield, viewed the space enterprise as a lower order of flying, one that was more robotic in nature and at odds with the test-pilot ethos. Jokes abounded that the astronauts were merely going along for the ride, no longer pilots but "spam in a can." Quietly but forcefully the astronauts compelled NASA to entertain key design changes in the spacecraft to allow a greater scope for the astronaut to exercise effective control over his spacecraft.[9]

SPACE AND THE NEW FRONTIER

The presidential election of 1960 proved to be a close contest. John F. Kennedy defeated Richard M. Nixon by just 118,000 votes—a mere 0.2 percent of the popular tally. The new president labeled his program the New Frontier, in a highly effective move to suggest that he offered vigorous new leadership for the United States. "Let every nation know, whether it wishes us well or ill," Kennedy asserted in his inaugural address, "that we shall pay any price, bear any burden, meet any hardship, support any friend, oppose any foe, in order to assure the survival and the success of liberty." These words calling for a forceful response to the challenges of the 1960s resonated well with the mood of the country. One legacy from the 1950s was the emerging space rivalry between the United States and the Soviet Union—and Kennedy soon found himself confronting this lingering issue from the Eisenhower years.

On the campaign trail in 1960, Kennedy had been blunt in his assessment of where the United States stood in the space race vis-à-vis the Soviet Union. "They [foreign nations] have seen the Soviet Union first in space," he told a crowd in Idaho. Referring to Russian space achievements, he noted: "They have seen it first around the moon, and first around the sun. . . . They have come to the conclusion that the Soviet tide is rising and ours is ebbing. I think it is up to us to reverse that point." Speaking at New York University in New York City's Washington Square in October, he asserted: "These are entirely new times, and they require new solutions. . . . The Soviet Union is now first in outer space."[10]

At another point in the campaign, Kennedy noted that American science and education enjoyed a reputation "second to none" around the world. Yet, he lamented, "[t]he first vehicle in space was called Sputnik, not Vanguard. The first country to place its national emblem on the moon was the Soviet Union, not the United States." Even the first dogs sent into space and safely returned to Earth "were named Strelka and Belka, not Rover or Fido, or even Checkers," Kennedy added.[11]

Kennedy's campaign rhetoric raised profound and nagging doubts about Eisenhower's leadership in the face of the Soviet challenge in space. Notwithstanding these charges, the Eisenhower administration could point to some substantial strides forward in the aftermath of the Sputnik launches. In fact, the relative standing of the United States and the Soviet Union in space was more nuanced than Kennedy's campaign rhetoric suggested. After a slow, even embarrassing start, the Eisenhower administration had made real progress in advancing the U.S. space program.

By October 4, 1960, the United States had orbited 26 satellites and launched two successful space probes, compared to six satellites and two space probes by the U.S.S.R. in the same period. It had responded to the Soviet space challenge on several levels, including development of new categories of satellites. One was Echo, a "passive" communications satellite in the form of a 100-foot diameter inflatable balloon that reflected radio waves back to Earth, permitting real-time two-way voice communication via space for the first time. Another important U.S. success was TIROS (Television and Infrared Observation Satellite Program), the world's first weather satellite, which transmitted thousands of images of cloud cover, severe storms, and other weather phenomena back to Earth. TIROS was the precursor of today's global meteorological satellite information system. Both were orbited in the final year of Eisenhower's term. The United States also achieved a significant technological breakthrough when its top-secret spy satellite program, Corona, successfully returned the first man-made object ever recovered from space.[12]

Another issue that loomed large in the 1960 campaign was the alleged "missile gap" between the United States and the Soviet Union. In the immediate aftermath of Sputnik 1, there had been the widespread fear that the Soviets would soon have hundreds of deployed ICBMs capable of delivering a catastrophic surprise thermonuclear attack on American cities. Khrushchev himself had fueled such speculation when he announced in the late 1950s that Russia was producing ICBMs "like sausages." On the campaign trail, Kennedy and his running mate, then U.S. Senate Majority Leader Lyndon Johnson, readily embraced the existence of this superpower missile gap. In August 1960, Kennedy, in a speech before the Veterans of Foreign Wars in Detroit, asserted that "the missile lag looms larger and larger ahead." A month later, he continued on the same theme, calling for "crash programs to provide ourselves with the ultimate weapons . . . which will eventually close the missile gap."[13]

The missile gap, in reality, was a myth, a fact obscured from the electorate in 1960. The Soviets had constructed only four R-7 ICBM launch pads, a small network with merely a half-dozen R-7s actually deployed.[14] But the true facts were only beginning to become known to U.S. intelligence agencies, given the very limited reliable knowledge available in the West about what was really taking place in the Soviet Union. Using the scant information available to them, the American intelligence community split between the Air Force analysts claiming that there could be as many as hundreds of deployed Soviet ICBMs and the CIA's view that there were no more than a dozen. These estimates changed frequently and analysts differed in their assessment of the actual threat.[15]

Beginning in 1956, U.S. efforts to pierce the Iron Curtain began to pay off. The desperate need to know definitively how many long-range ballistic missiles were in the Soviet arsenal became the driving force behind two major American strategic intelligence programs: the U-2 spy plane and the Corona photo-reconnaissance satellite. The U-2 overflights, initiated in 1956, covered only part of the vast Soviet land mass. The U-2 intelligence data suggested no large Soviet ICBM deployment, but the Air Force seized on the incomplete coverage of the U-2 airborne cameras to press the argument that large numbers of ICBMs might still lurk unseen, arguing that it remained prudent to deploy a large number of American ICBMs. Coincidentally, the only American ICBM ready for deployment at that time was the Air Force's own Atlas. When Kennedy made the "missile gap" a campaign issue in 1960, many in the public sphere expressed alarm and a sense of urgency over national security.[16]

This claim was made despite the fact that both Kennedy and his running mate, Lyndon Johnson, had been briefed by then CIA director, Allen Dulles, on the exact nature of Soviet ICBM deployment. The briefing took place in July 1960, a session where Dulles—in his own words—provided "an analysis of Soviet strategic attack capabilities in missiles and long-range bombers." Notwithstanding the intelligence briefing, Kennedy chose to maintain the "missile gap" as an effective campaign issue in the last weeks of the 1960 presidential campaign.[17]

The dramatic series of Soviet space triumphs—all launched by the powerful R-7 rocket—added credibility to this bogus issue. During the campaign Nixon winced at the charge that the Soviet triumphs in space dwarfed the American effort. He took issue with Kennedy's critical comments, asserting that it was "irresponsibility of the highest sort for an American presidential candidate to obscure the truth about America's magnificent achievement in space in an attempt to win votes."[18] Nonetheless, the Soviet Union did enjoy some significant advantages over the United States, which Kennedy sought to exploit in his campaign rhetoric. Primary among these was Russia's ability to launch into orbit vehicles weighing far more than U.S. spacecraft, including the five-ton Vostok, launched in May 1960, a test vehicle later used for Russia's manned orbital flights. As it turned out, the Vostok, complete with a "dummy" astronaut, was intended to return to Earth, but a malfunction sent it instead into a higher orbit.

On January 19, the day before John F. Kennedy's inauguration, Glennan spent his last working day at NASA, departing Washington in a winter blizzard for his home in Ohio. Glennan's departure offered the Kennedy administration the option of appointing a new administrator, a person who would embrace fully the space program of the New Frontier. The man selected for the job was James E. Webb, who assumed his duties in February 1961. Abe Zarem, a well-respected scientist and engineer, felt Webb fit well the demanding portfolio for the space agency—

"an evangelist, with a keen sense of our national rendezvous with destiny . . . an efficient manager . . . a man of exceptional social manners, particularly for briefing Congress."[19] In addition to his extraordinary energy and strong organizational skills, Webb brought a wide variety of experience to his new position. Trained as a lawyer, he had been a highly successful business executive at Kerr-McGee Oil in Oklahoma. His ties to Washington were substantial: He had been a congressional aide, Truman's director of the budget, and undersecretary of state under Dean Acheson.[20] Webb always sought out talented people for his inner circle, a trait that became evident immediately when he chose Hugh Dryden as his deputy administrator, a man who had served NASA under Glennan. Though Webb lacked technical training, he approached the leadership of NASA with boldness and vision.[21]

Jack Valenti, then a Houston newsman and a future aide to President Lyndon Johnson, wrote a brief biographical sketch of Webb, titled "Thank the Lord for the Good Men." Valenti described Webb as a dynamic and forceful leader of the new space agency, moving with "the energy of his Atlas boosters." Those who worked at NASA headquarters with Webb also made a positive impression on Valenti, in particular Robert Seamans, Hugh Dryden, and Robert Gilruth. In Valenti's words: "They have found rapport with achievement. They have built a strong house where valor dwells."[22]

Webb assumed the leadership of NASA at a time when it was expanding through a variety of new program initiatives. In 1961, four new offices were established: manned spaceflight, space sciences, applications, and advanced research and technology. By the end of that year, NASA had just under 18,500 full-time civilian employees. In addition, its contractors employed 58,000 people in 1961 and 116,000 the following year. This was just the start of an expansion that would peak at 377,000 jobs in 1965. NASA's budget grew apace from 964 million dollars in 1961 to a total of 32 billion dollars for its first 10 years. As a sign of that growth to come, NASA facilities grew as well. First to be addressed was the need for a separate operation for manned spaceflight operations. Following a review of 20 cities, Webb announced in September 1961 that Houston would be the site of a facility that would design, develop, and manufacture all manned spacecraft; select and train their crews; and oversee their actual space missions. Texas was the home state of Vice President Johnson, a strong supporter and initiator of the space program. In 1962, NASA also took over 111,000 square acres at Cape Canaveral, next to the military launch sites there. The Department of Defense had run the entire facility from May 1949, when President Truman had authorized the missile launch range. In addition, the various NASA centers were supplemented by a rocket-engine test facility in Mississippi and an electronics research center in Massachusetts, among others.[23]

There was one fitting, if belated, move by the new Kennedy administration to correct the hyperbole of the 1960 campaign regarding missiles. In January 1961, just days after his inauguration, Kennedy ordered his new defense secretary, Robert McNamara, to conduct a full review of the Missile Gap issue. The results of that study were summed-up in the headline on a page one *New York Times* story published on February 7, 1961: "Kennedy Defense Study Finds No Evidence of a Missile Gap." The piece set the historical record straight, reporting "studies made by the Kennedy administration since Inauguration Day show tentatively that no 'missile gap' exists in favor of the Soviet Union. The conclusion appears to back the views of former President Eisenhower, who told Congress last month that the missile gap 'shows every sign' of being a fiction."

For all his pronouncements, Kennedy entered the White House with little knowledge or interest in space, except for his keen appreciation of how issues involving the American space program dovetailed with the politics of the Cold War. He was not a visionary or, for that matter, necessarily enraptured with the romantic prospects of space travel. Kennedy shared the existing American consensus on the Cold War, the necessity of advancing the national security of the United States in the era of nuclear weapons. In diplomacy, he expressed a keen sense of realpolitik and worked hard to maintain the balance of power and spheres of influence in American-Soviet relations.[24]

Hugh Sidey, the White House correspondent for *Time* and *Life* magazines, enjoyed extraordinary access to Kennedy. He wrote, "Of all the areas of bafflement when Kennedy took office, space seemed more perplexing than the others. Kennedy seemed to know less about it, be less interested in it."[25] However, Kennedy was committed to supporting a more robust space program than his predecessor. And he envisioned increased emphasis on manned spaceflight as opposed to the narrow agenda of launching communications, mapping, and weather satellites. This focus ran parallel to his robust support for the American military space program, including ongoing ICBM development, the Navy's submarine-launched Polaris missile program, and reconnaissance satellites. As president, Kennedy recognized the strong public appeal of the Project Mercury astronauts then in training and wanted to ensure that the program received full support from his administration.[26]

Even before Kennedy assumed office, NASA had achieved the important goal of taking control of the scattered institutions essential for a viable space program—part of the legacy of the Glennan years. NASA acquired the Navy's Project Vanguard staff in November 1958. As a civilian agency, NASA also acquired the U.S. Army's two prized space-age possessions—the Jet Propulsion Laboratory in Pasadena and the von Braun rocket team at the Redstone arsenal. (The Army, in its quest for a continuing and even expanded space role, had consolidated its Ballistic Missile Agency, JPL, and other related agencies under General John B. Medaris, in early 1958, creating the Army Ordnance Missile Command.) The process of consolidation was met with resistance and no small amount of infighting. For example, Medaris quickly acquiesced to the transfer of JPL, but resisted fiercely the transfer of the army's missile programs, a struggle in which he was actively supported by von Braun and his team. Yet the NASA takeover of Medaris' empire was complete by July 1960, when the Redstone Arsenal was renamed as the George C. Marshall Space Flight Center. With von Braun and his team came yet another great prize for NASA—the design work for the mighty Saturn rocket, which would later power the Apollo missions.[27]

By January 1961, NASA was ready to send into space a primate stand-in for the astronauts to come, using a Redstone, the same missile the astronauts would use on suborbital flights. (In May 1959, two monkeys,

Able and Baker, had survived a suborbital flight in the nose cone of a Jupiter IRBM.) Ham, a chimpanzee, was selected for this honor, and his 17-minute suborbital flight and recovery on January 31 were successful. The choice did not please the Mercury astronauts, who were chagrined by the idea of sending primates into space before a man. Two of them, Slayton and Shepard, later wrote, "The irony of playing second fiddle to a chimpanzee was particularly galling to these highly intelligent and skilled men."[28] Soon enough, though, events would renew the focus of the astronauts and the nation on the task of getting Americans into space.

A RED STAR

The Soviet space program, Tom Wolfe aptly observed in *The Right Stuff*, maintained "an aura of sorcery." The Soviets, he noted, "released practically no figures, pictures, or diagrams. And no names; it was revealed only that the Soviet program was guided by a mysterious individual known as the Chief Designer. But his powers were indisputable! Every time the United States announced a great space experiment, the Chief Designer accomplished it first, in the most startling fashion."[29] The year 1961 would offer little respite for Americans—indeed, the Chief Designer had some more surprises in store for NASA and the world.

The new arena for competition became manned spaceflight, the urgent quest to launch a human being into orbit—at the time still a radical notion. Project Mercury looked ahead to this seminal moment. The Soviets, even if operating under a shroud of secrecy, also hinted at such a bold undertaking. In fact, Korolev and his associates had made substantial progress testing prototype capsule-satellite technologies. The Korabl-Sputnik series, dating back to May 1960, used dogs and other animal species to test the impact of spaceflight on living creatures. Some of the Korabl-Sputnik missions failed, typically in the reentry phase, killing the animals; others, such as two launches in March 1961, proved to be highly successful. These same experimental launches offered an avenue for perfecting landing techniques of capsules after reentry into Earth's atmosphere. Upon reaching a safe altitude, the canines were ejected from the capsule and parachuted to Earth, the very same system that would be used in future missions with cosmonauts. These tests became part of what later was known as the Vostok series.[30]

Two grim footnotes to the triumphal march of the Soviet program occurred during the interregnum between Sputnik and the debut of manned spaceflight: the explosion of the R-16 ICBM. On October 24, 1960, the prototype R-16 blew up on the launch pad after a fire erupted in the second stage of the rocket. The massive conflagration that followed killed about 130 people, including General Mitrofan I. Nedelin, then commander of the Soviet Union's Strategic Rocket Forces. Test pilot Dolgov met his death in an accident while testing the ejection seat mechanism for a future manned spacecraft. These setbacks remained concealed behind a curtain of secrecy—only the triumphs of the Soviet space program were trumpeted to the West.[31]

As with Project Mercury, the Soviets had quietly begun the arduous process of screening candidates for their cosmonaut corps. The preparatory work was broadly based and in some respects parallel in its character to the American program. The Soviet Air Force, for example, established a special department in aviation medicine to focus exclusively on the new realm of space.[32] By March 1961, this effort had reached maturation, and a group of finalists was introduced to the emerging Vostok spacecraft technology. At this juncture, in anticipation of an imminent manned spaceflight, cosmonaut training proceeded at a rapid pace. No less important, Soviet technicians worked feverishly to perfect the life-support system for the capsule, the space suit, and the critical ejection-seat mechanism. The preparatory phase for manned orbital flight reached a critical milestone on April 8, 1961, when Yuri Gagarin, one of six finalists, was designated as pilot, and German Titov received the nod to serve as his backup pilot. Three days later, both men met with engineers and technical staff at the launch pad for a final briefing. The date for the mission was now set for April 12. The Vostok capsule would be launched at 9:07 A.M. (Moscow time). Korolev remained at the epicenter of these events, working with his talented staff, including Nikolai Kamanin, in charge of cosmonaut training, Mstislav V. Keldysh, a distinguished mathematician and physicist, and Konstantin Feoktistov, one of the leading engineers associated with the Vostok program.[33]

Yuri Gagarin was 26 years old when he joined the elite group of cosmonauts in January 1961. Before that time, Gagarin and his fellow cosmonauts had endured a rigorous period of preparatory work in simulators and parachute training. Gagarin came from the Smolensk region, west of Moscow, from a family with a proletarian pedigree. Having graduated from secondary school, he entered the Soviet Air Force, qualifying as a military pilot at the Orenburg Higher Air Force School in 1955. Assigned to an airfield at Zapolyarniy, in the far north above the Arctic Circle, he quickly established himself as a talented aviator, but he never became a test pilot like his American counterparts in the Mercury program.

Once chosen as a candidate for the cosmonaut program, Gagarin—with his engaging smile—made a positive impression on all with his intelligence, motivation, and discipline. B. V. Raushenbakh, a close associate of Korolev in OKB-1, remembered Gagarin for his innate modesty and tact, his impressive memory and attention to detail, his quick responses, and his aptitude for celestial mechanics and mathematics. Gagarin could be forceful and outspoken on matters of principle, but his manifest talents were never compromised by arrogance or undue self-promotion. He was popular and highly respected by his fellow cosmonaut trainees. Korolev was impressed with Gagarin when he met with the six finalists for the first time; some believed this interview elevated the young pilot from Smolensk to the forefront for the pivotal Vostok mission. One final consideration, if unstated, was the fact that Gagarin came from an ethnic Russian background (Great Russian). This favored ethnicity, combined with his impeccable working-class background, confirmed Gagarin's eligibility to be the first man into space.[34]

On the eve of the historic flight, Gagarin and his backup pilot, German Titov, were housed in a special cottage near the launch pad at Baikonur. Doctors attached sensors to both cosmonauts to monitor their vital signs. They slept well that night, notwithstanding the excitement associated with the launch. By 5:00 A.M., even before Gagarin awoke, the various ground stations tested their communications links. As these tests went forward, both cosmonauts were awakened at 5:30 A.M. Korolev himself had spent a sleepless night, fretting over the potential problems that might arise with the third stage of the Vostok spacecraft—what if it failed in the ascent phase, forcing an emergency descent in the ocean near Cape Horn?

The fateful moment finally arrived: Gagarin, in his space suit, was driven to the launch pad. As a precaution, Titov was dressed in the same way, in case of a last-minute requirement to replace Gagarin. Given the significance of the event, Gagarin was greeted by Konstantin N. Rudnev, who headed the State Commission for Vostok. The pace then quickened to meet the scheduled time for liftoff at shortly after 9:00 A.M. Korolev oversaw the seating of Gagarin in the launch vehicle, and his technicians attended to countless details in the launch-preparation sequence. When the hatch was closed, it was discovered that one of the sensors would not function, so the hatch had to be reopened, the sensor adjusted, and the hatch closed again. At T minus 15 minutes, Gagarin put on his sealed gloves and helmet. The tower arms moved away. As the tension mounted, Korolev took a tranquilizer pill to calm his nerves. He and his staff knew that R-7—if a powerful rocket—did have a large number of launch mishaps.[35]

Atop the R-7, Gagarin waited anxiously for the liftoff. T minus five minutes. T minus one minute. Lift off came at exactly 9:06 hours, 59.7 seconds (Moscow time).[36] Gagarin's pulse had reached 157 beats per second just before the engines of the R-7 roared to life. As he ascended into the sky, Gagarin yelled, "We're off!"

At 09:09 hours: Korolev spoke to his cosmonaut.

Korolev: "T plus 100 [seconds]. How do you feel?"

Gagarin: "I feel fine. How about you?"

The rocket gained momentum as the seconds passed, pushing Gagarin into the thin upper atmosphere. He felt initially the pressure of five *g*'s. The muscles on his face were strained. He reported that he encountered increased difficulty in talking normally. As scheduled, the nose fairing on the rocket separated. The main core stage and strap-on rockets then detached and fell away. Once free, the upper stage of the R-7 rocket ignited and propelled Gagarin into orbit.

The communications with Korolev resumed at 09:10 hours.

Korolev: ". . . How do you feel?"

Gagarin: ". . . nose fairing jettisoned. . . . I see the Earth. The g-load is increasing somewhat. I feel excellent, in a good mood."

Korolev: "Good boy! Excellent! Everything is going well."

Gagarin: "I see the clouds. The landing site . . . it is beautiful. What beauty! How do you read me?"

Korolev: "We read you well, continue the flight!"

Once Gagarin reached orbit, he traveled at a speed of 17,500 miles per hour. He looked out on a remarkable scene, a celestial view of Earth from the stupendous height of over 100 miles; the Vostok orbit ranged from a low point of 108.5 miles to 187.2 miles. "The Earth," Gagarin recalled, "began to pass to the left and up, then to the right and down. . . . I could see the horizon, the stars. . . . The sky was completely black, black. The magnitude of the stars and their brightness were a little clearer against the black background. . . . At the very surface of the Earth, a delicate light blue gradually darkens and changes into a violet hue that steadily changes to black."[37]

Gagarin reported to mission control on some of the more mundane details inside the capsule. He discovered that he could eat and drink. (He was the first human to have a meal in orbit.) He was amazed with the altered state of weightlessness. His tablet and pencil floated freely in the cramped interior of the spacecraft. His pencil moved slowly out of reach. He reported no negative consequences or unpleasant sensations. "Here," Gagarin observed, "you feel as if you were hanging in a horizontal position in straps."[38]

The Russian cosmonaut maintained contact with Earth through high-frequency radio and by using a telegraph link. The American CIA and the National Security Agency, using an electronic intelligence station in Alaska, were able to intercept the Vostok's radio and television transmissions. The TV images confirmed that there was a man on board, not a dummy.[39] The Soviet news agency TASS did not report the Gagarin flight for an hour, a belated confirmation of the launch with a passing reference to the cosmonaut on board: "The pilot-cosmonaut of the spaceship satellite Vostok is a citizen of the Union of Soviet Socialist Republics, Major of Aviation Yuri Alekseyevich Gagarin."[40]

The flight plan called for just one orbit, which precluded any elaborate experiments in a short time frame of 108 minutes. At this point in space exploration, no extensive data on the effects of prolonged weightlessness on humans had been recorded; consequently, the Soviet medical team remained concerned about the effect of zero gravity on Gagarin's body and psyche. Given all these uncertainties, Korolev had designed the Vostok as a highly automated vehicle, with only a minimal role for the pilot. Gagarin was a guinea pig in this formative moment in the space age, but he did have on board a special envelope with the key code to unlock the spacecraft's controls in an extreme emergency.

The descent to Earth presented unforeseen perils for Gagarin. When the retro-rockets in his instrument module fired for 40 seconds, a precise maneuver to slow the spacecraft for reentry, the craft experienced a sharp jolt, prompting the Vostok to enter a spin at high speed. The *g*-loads increased, as did the heat generated by the reentry. The instrument module had been programmed to separate at 10 seconds after retrofire, but this critical maneuver was delayed momentarily. When this malfunction occurred, the Vostok was over Africa. Caught in the violent spin, Gagarin gazed anxiously out the porthole as the African continent filled the window, then Earth's horizon, and then the black void of space. "Suddenly a bright purple light," Gagarin later reported, "appeared at the blind edges. . . . I felt oscillations of the spaceship and burning of the coating [thermal protection layers]. . . . I heard crackling sounds. . . . By the time the load factor reached its peak, the spaceship oscillations reduced to 15 degrees. At that moment I felt that the load factor reached about 10 *g*'s. There was a moment for about two or three seconds, when the instrument readings became blurred. My vision became somewhat grayish. I strained myself again. This worked, and everything assumed their proper places."[41]

Finally, the Vostok righted itself, entering a descent into the upper atmosphere at a proper angle. Throughout the dangerous passage Gagarin had kept his cool, reporting regularly on his condition and the status of the spacecraft. When the spacecraft descended to an altitude of 23,000 feet, the main parachute opened, breaking the fall to Earth. In seconds, the Vostok's hatch blew open. Gagarin was thrust outward: "I'm sitting there thinking that wasn't me that was ejected, was it? Then I calmly turned my head upward, and at that moment, the firing occurred, and I was ejected. It happened quickly, and went . . . without a hitch. I did not hit anything. . . . I flew out in the seat."[42]

He opened his own parachute around 13,000 feet, making a soft landing in a field near a ravine outside the city of Saratov near the Volga River. Once disentangled from his parachute harness, Gagarin climbed a small hill, where he encountered a woman and a small girl. He rushed forward, waving his hands and shouting, "I'm a friend, I'm Soviet!" Learning of the spaceman's urgent desire to contact his controllers, the woman invited him to use the telephone in the nearby field camp. Gagarin later observed: "I told the woman not to let anyone touch my parachute while I was going to the camp. As we approached the parachutes, we saw a group of men, about six all in all—tractor drivers and mechanics from the field camp. I got acquainted with them. I told them who I was. They said that news of the spaceflight was being transmitted at that moment over the radio."[43]

Pravda, the official news organ of the Communist Party, gave the Gagarin feat banner headlines on April 13, announcing the flight as a "Great Event in the History of Humanity." The news was tinged with an explicit political spin, proclaiming that the "space first" had been achieved under the "Banner of Lenin." The headlines were addressed "To the Communist Party" and "To All Progressive Humanity." Two days after the launch Yuri Gagarin arrived in Moscow in a military transport escorted by seven fighter jets. Nikita Khrushchev proclaimed that Gagarin's feat had made him "immortal," awarding him the Gold Star medal, "Hero of the Soviet Union," the highest honor for bravery. The young cosmonaut became the object of intense media attention and popular enthusiasm. Boris Chertok in his memoirs compared this outpouring of patriotism to Victory Day at the end of World War II.[44]

The mysterious Chief Designer had scored another dramatic victory over the United States.

JOHN F. KENNEDY OFFERS A CHALLENGE

The specter of another Soviet space triumph haunted Kennedy and his advisers. The new president had received regular intelligence reports that the Soviets might attempt to launch the first man in space. The reports became serious enough that presidential press secretary Pierre Salinger and his press colleagues at the Defense and State Departments began drafting a presidential statement for use if the Soviets succeeded. On the night of April 11, after the president had tentatively signed off on the statement, his science adviser, Jerome Wiesner, stopped by the Oval Office to tell Kennedy that American intelligence felt there was a strong chance the Soviet spaceflight would take place that night. Later, Major General Chester Clifton, the president's military aide, checked in with the president for the final time that day. "Do you want to be waked up?" he asked. "No, give me the news in the morning," Kennedy replied. Just hours later, American intelligence received the first intercepts indicating that Yuri Gagarin had launched off the pad at Baikonur and headed for orbit. The president was not disturbed and only learned about the flight the next morning.[45]

The Soviets posed Kennedy with a major new challenge, ironically in the aftermath of a presidential election where he had repeatedly asserted the need for America to do more in space. He had not placed space at the top of his priorities in the opening months of his presidency. For example, in March, James Webb, his new NASA administrator, had asked for additional funding for both the Saturn rocket and the nascent Apollo project, which at the time had as its long-term goal men flying around the moon rather than landing on it. Webb proposed to greatly accelerate and expand the Apollo project to land men on the lunar surface by the end of the decade. Kennedy considered the proposal, which was strongly opposed by his budget bureau director on grounds of its high cost, and rejected it. Instead, the president agreed to more funding for the Saturn only.[46]

After the orbital flight of Gagarin, the space race moved to the forefront of the president's agenda. On April 14, Kennedy summoned key officials of his new government to the White House to discuss the public perception that America was losing the space race. At the time, White House correspondent Sidey was working on a story on the American lag in the space race, and the Gagarin flight suddenly gave his story great relevance. To Sidey's delight and surprise, presidential aide Ted Sorensen invited him to sit in on Kennedy's meeting, providing a unique window on the president's thinking at this crucial juncture for his administration. Others attending the session in the cabinet room included Webb and his deputy, Hugh Dryden, budget director David Bell, and science advisor Wiesner. Despite the presence of the reporter, frank and revealing comments were offered as the president asked the men for their thoughts on what the United States should do in response to the latest Russian space triumph. After listening to his experts, the president frowned and observed, "We may never catch up." A moment later, he asked, "Is there any place where we can catch them? What can we do? Can we go around the moon before them? Can we put a man on the moon before them? . . . Can we leapfrog?"[47]

Dryden commented that only an effort on the scale of the Manhattan Project could get the U.S. to the moon, an effort that might cost 40 billion dollars. Even with that effort, he added, they had only a 50 percent chance of beating the Soviets there. "The cost, that's what gets me," Kennedy said, looking over at Bell, who responded that the cost of scientific ventures rose in geometric proportions. Following some additional comments by other participants, the president summed up his feelings with astonishing candor, and even desperation, considering that a reporter was present: "When we know more, I can decide if it's worth it or not. If somebody can just tell me how to catch up. Let's find somebody—anybody. I don't care if it's the janitor over there, if he knows how." After a searching look at the faces around him, the president added, "There's nothing more important."[48] In a 1971 recounting of the meeting, Sidey provided additional details: "Kennedy was very anguished during the meeting. He . . . slumped down in his chair. . . . He kept running his hands through his hair, tapping his front teeth with his fingernails, a familiar nervous gesture." An interesting postscript to Sidey's account of the meeting was contained in a 1979 article he wrote on the 10th anniversary of the successful Apollo 11 moon landing. Sidey added a detail not mentioned in his earlier articles or his book on Kennedy: After the meeting was over, Sorensen went back and talked briefly with the president. Sorensen then emerged to tell Sidey, "We are going to the moon."[49]

On April 17, exactly a week after Gagarin's flight, a second major setback for the administration would play a critical role in pushing Kennedy toward a decisive commitment to land Americans on the moon. On that date a group of Cuban refugees stormed ashore at the Bay of Pigs on the Caribbean island in an American-sponsored attempt to foment a popular revolution against Communist leader Fidel Castro. Denied promised U.S. air support, and finding little support from the Cuban people, the invasion was a fiasco from the start. U.S. backing of the invasion greatly embarrassed Kennedy and his administration and damaged relations with other nations. While the debacle was not cited explicitly as a reason for the eventual Apollo go-ahead, Kennedy clearly sought a new initiative to help restore the nation's tattered prestige.[50]

Several days after the cabinet room meeting, Kennedy asked Vice President Johnson for his suggestions on the next steps for America's space program. Johnson chaired the National Aeronautics and Space Council, an entity originally established in the legislation creating NASA. The council operated as part of the executive office of the president, and its members included, in addition to the vice president, the secretaries of state and defense, and the NASA administrator.[51] Johnson suggested that the council should consider the issue and recommend a space program that would be supported by Congress and the public.

Although Johnson had shown little interest in space exploration before Sputnik 1 was orbited, he quickly seized upon the issue. Three months later, he said in a speech that "control of space means control of the world . . . [t]hat is the ultimate position, the position of total control over the Earth that lies somewhere in outer space."[52] After Johnson became vice president, he remained a staunch advocate of space exploration and the massive appropriations required to sustain it.

The next day, April 20, 1961, Kennedy sent Johnson a memorandum asking, "Do we have any chance of beating the Soviets by putting a laboratory in space, or by a trip around the moon, or by a rocket to land on the moon, or by a rocket to go to the moon and back with a man. Is there any other space program which promises dramatic results in which we could win?"[53]

Now animated with a sense of urgency, Kennedy asked that a Space Council report be given to him "at the earliest possible moment." While awaiting the response, Kennedy made a public comment that clearly reflected his thinking at the time. "If we can get to the moon before the Russians, then we should," he told a news conference on April 21, also mentioning his request to Johnson to review U.S. space options. The president said nothing further in public about sending Americans to the moon before he announced Project Apollo five weeks later.[54]

As the Space Council began considering the options available, it reviewed NASA's original proposal for a manned circumlunar flight by 1966-1968 and a manned lunar landing sometime after 1970. Both goals were included in NASA's 10-year plan, developed at the request of Congress. Eisenhower, always ambivalent about spending huge sums on what he saw as a pointless space race with the Soviets, had refused to sign off on what he regarded as an overreaching plan to send men to the moon. At one point, he commented that he had no intention of hocking the family jewels "like Isabella" did for Columbus' voyage to the New World, in order to finance a U.S. manned lunar mission. Regardless of the president's views, NASA under Glennan had continued to develop plans for a manned space program. In July 1960, 1,300 government, aerospace industry, and academic institution representatives attended a NASA-sponsored conference in Washington to discuss advanced manned spaceflight programs, including circumlunar voyages and eventual manned lunar landings. That same month, the name "Apollo" was approved for the advanced manned flight program.[55] Once again, preparatory work done by the former administration provided a strong base for the expansion of NASA in the early 1960s.

On April 28, in a six-page interim report to President Kennedy, Vice President Johnson observed, "The U.S. can, if it will, firm up its objectives and employ its resources with a reasonable chance of attaining world leadership in space during this decade." He went on to discuss the practical requirements: "This will be difficult but can be made probable even recognizing the head start of the Soviets and the likelihood that they will continue to move forward with impressive successes. . . . If we do not make the strong effort now, the time will soon be reached when the margin of control over space and over men's minds through space accomplishments will have swung so far over to the Russian side that we will not be able to catch up, let alone assume leadership. . . ." He urged Kennedy to assume the initiative, to appreciate the propaganda value of spaceflight, and to pave the way for the United States to forge ahead with a "manned exploration of the moon."[56]

Kennedy's ultimate decision to commit the United States to a lunar mission must be viewed through the prism of the rivalry between the two superpowers. The intense competition engendered a strongly felt American need to respond to Soviet space triumphs by showing that America was at least as capable as the Soviets in this area. As a strong advocate of a manned lunar mission, Vice President Johnson encouraged others on the Space Council to follow suit. One result of his efforts was a top-secret memo signed by Defense Secretary McNamara and NASA administrator James Webb in May 1961. "Major [space] successes," they argued, "such as orbiting a man as the Soviets have just done, lend national prestige even though the scientific, commercial or military value of the undertaking may by ordinary standards be marginal or even economically unjustified." But, they asserted, in the case of going to the moon: "Our [space] attainments are a major element in the international competition between the Soviet system and our own. The non-military, non-commercial, non-scientific but 'civilian' projects such as lunar and planetary exploration are, in this sense, part of the battle along the fluid front of the Cold War."[57]

The clear import of the McNamara-Webb memorandum was that the American space program, whether in military or civilian form, was an essential component in winning the Cold War. By early May, Kennedy National Security Adviser McGeorge Bundy would later recall, the president essentially had made the decision to go ahead with Project Apollo and had little interest in hearing arguments to the contrary.[58]

President Kennedy addressed a joint session of Congress on May 25, 1961. He delivered what he characterized as a second State of the Union Address for the year, made necessary by "the extraordinary times" in which the nation found itself. The address contained a space component, fueled by the recent Gagarin flight and the entire panoply of past Soviet space accomplishments: "Finally, if we are to win the battle that is now going on around the world between freedom and tyranny, the dramatic achievements in space which occurred in recent weeks should have made clear to us all, as did Sputnik in 1957, the impact of this adventure on the minds of men everywhere who are attempting to make a determination of which road they should take. . . . Now it is time to take longer strides— time for a great new American enterprise—time for this nation to take a clearly leading role in space achievement which, in many ways, may hold the key to our future on Earth." Citing the time factor and the Soviet Union's significant lead in large rocket engines, Kennedy observed, "We nevertheless are required to make new efforts on our own. For while we cannot guarantee that we shall one day be first, we can guarantee that any failure to make this effort will make us last."

Kennedy then concluded with his oft-quoted challenge to countrymen, "I therefore ask this Congress, above and beyond the increases I have earlier requested for space activity, to provide the funds which are needed to meet the following national goals: First, I believe that this nation should commit itself to achieving the goal, before this decade is out, of landing a man on the moon and returning him safely to Earth. No single space project in this period will be more impressive to mankind and more important for the long-range exploration of space. And none will be more difficult or expensive to accomplish."[59]

Despite the ringing declarations of Kennedy's May 1961 speech, he continued in private meetings to voice his lack of real interest in space. At the same time, he acknowledged candidly the primary role of Cold War considerations in driving U.S. space spending and policies. On November 20, 1962, 18 months after his call for an American manned lunar landing, Kennedy met with NASA administrator Webb and budget bureau director Bell to discuss concerns that NASA was not paying sufficient attention to Project Apollo and that additional funding might be required for the human spaceflight program. While the meeting was recorded, the tape and a transcript were not made public until August 2001. Two comments by the president provided dramatic insight into his true feelings. "Now, this may not change anything about that schedule but at least we ought to be clear, otherwise we shouldn't be spending this kind of money because I'm not that interested in space," Kennedy said. Just a minute earlier, he had said, "And the second point is the fact that the Soviet Union has made this a test of the system. So that's why we're doing it."[60]

THE SUBORBITAL FLIGHTS

Even before Kennedy announced the nation's commitment to send men to the moon, America, like a proverbial sleeping giant, had already awakened to its coming role in space. On May 5, Project Mercury sent the first American, astronaut Alan Shepard, into space, albeit only for 15 minutes on a suborbital ride to an altitude of 116.5 miles.

A Navy pilot, Shepard had flown some of the latest jet fighters at the Patuxent River Naval Air Station, Maryland, and belonged to an elite circle of senior Navy test pilots. He had also performed aircraft carrier-landing suitability and in-flight refueling trials for new Navy fighters, and served two tours with carrier-based squadrons.

Shepard's flight was just what the nation needed at that point, and 45 million Americans watched on television as Shepard's Freedom 7 Mercury capsule lifted off from its Cape Canaveral launch pad atop a Redstone missile. Among them was the president, watching at the White House along with First Lady Jackie Kennedy and senior administration officials, including Lyndon Johnson. Only after Shepard's liftoff and a tense wait for word that he had been recovered and was onboard a rescue helicopter did the president smile and relax. He turned to the others and said quietly, "It's a success."[61]

The suborbital jaunt proved to be a success with no major mishap, and Shepard's impatience to go into space was later immortalized in Tom Wolfe's *The Right Stuff.* With less than three minutes remaining in the countdown, Mercury control called a hold, not the first one that morning. The prospect that the hold could possibly result in a postponement did not make Shepard happy. "All right," he snapped. "I'm cooler than you are. Why don't you fix your little problem . . . and light this candle." That seemed to work; a few minutes later, Mercury Control declared the launch a "go," and Shepard was on his way.[62] Shepard encountered some buffeting at the transonic zone, which turned more severe when the spacecraft reached maximum dynamic pressure at 90 seconds. At this point the buffeting became rather extreme, bouncing Shepard's head so violently he could not read his instruments. At the two-minute mark, the astronaut experienced the maximum *g* forces. Then 22 seconds later, the engines cut off; at this point Shepard was traveling at a speed of 5,134 mph. Initially, he had flown face-forward, but now for reentry, the capsule turned around automatically. Shepard tilted the heat shield downward 34 degrees for the passage through the upper atmosphere. When the retro-rockets fired to slow the spacecraft, it jolted the astronaut with what was later called a "comforting kick in the ass."[63]

The first American suborbital flight was executed in full public view from liftoff to recovery, in sharp contrast to the near-total secrecy attending Gagarin's flight a few weeks earlier. Shepard's short trip was not lacking in technological significance, either. Nearing the top of his suborbital ride, Shepard switched Freedom 7's attitude-control system from automatic to manual, one axis at a time: pitch (up and down motion), yaw (left and right motion), and roll (revolving motions). He experimented briefly with all three axes, a key exercise in demonstrating the importance of pilot control by the astronauts.[64] Predictably, Shepard was embraced by the American public as a genuine hero, and welcomed to Washington, D.C., after his flight with a huge parade down Pennsylvania Avenue attended by more than 250,000 people.

The second manned Mercury suborbital flight took place on July 21, 1961. Virgil "Gus" T. Grissom took command of the spacecraft that he had dubbed Liberty Bell 7. Grissom, an Indiana native, brought to the Mercury program a stellar record of achievement. As an Air Force pilot he had flown 100 combat missions in Korea at the controls of the fabled F-86 Sabre jet fighter. He had won the Distinguished Flying Cross for chasing off a North Korean MiG-15 seeking to shoot down an American reconnaissance aircraft. He went on to become a test pilot, checking out all-weather fighters for the Air Force. For Grissom's short flight, providing 10 minutes of weightlessness, the focus shifted to as much visual observation as possible. The capsule had one window, which allowed for nearly 30 degrees of forward vision—up, down, and sideways. He lifted off at 7:20 A.M. Having reached an altitude of 118 miles, he repeated some of the same manual-control tests as Shepard and then quickly prepared for reentry. All went well in the descent phase. Once the capsule touched down in the Atlantic Ocean, Grissom prepared for his recovery from the capsule, radioing two nearby recovery helicopters to come and pick him up.

Liberty Bell 7 had a side escape hatch cover that was secured by explosive bolts; the astronaut was to remove a lock pin and then press a plunger in order to blow the door free and allow his safe exit from the capsule. A total of 70 bolts, each fitted with a detonation fuse, secured the side hatch. When the astronaut pulled the pin in the cockpit, a fist force of five or six pounds blew open the hatch. Grissom alerted the rescue helicopter. As the rescue party approached, a near disaster unfolded: "I was lying there, minding my own business when I heard a dull thud," Grissom later reported. Suddenly, the hatch cover blew away and salt water began filling the capsule. Grissom told a board of inquiry later that while he was unable to recall all of his actions at this point, he was certain that he had not touched the hatch-activation plunger. Grissom emerged from the capsule and swam away.

The helicopter pilots assumed that Grissom could take care of himself in the water, and one helicopter began an attempt to recover the capsule. Grissom nearly drowned. As the capsule began to submerge, he was barely able to grab the collar dropped by the helicopter and had to be pulled to safety. The valve on his space suit had not been closed, so it quickly filled with water. Grissom was pulled upward to the awaiting rescue helicopter. In the end, the water-filled capsule proved too heavy for lifting by the helicopter and had to be cut loose, sinking to the ocean floor, some 2,800 fathoms under the Atlantic, not to be retrieved for 40 years.[65] Grissom's saga was a grim reminder of the inherent dangers of reentry and rescue. Both suborbital flights, notwithstanding the loss of Liberty Bell 7, had been successful. Still, they appeared anemic and modest when compared to the achievement of Yuri Gagarin.

TITOV SCORES ANOTHER SOVIET VICTORY

The extraordinary success of Gagarin's spaceflight set the stage for a second Soviet manned feat—the dramatic launch of German Titov, Gagarin's backup, for an extended 17 orbits of the Earth. For the first time a human being would be in space for an entire day. The date for the space venture was set for August 1961, coming in the aftermath of America's suborbital flights. If successful, the Titov mission would clearly demonstrate the Soviet Union's superiority in manned spaceflight. The decision had not come without controversy. Several prominent physicians associated with space medicine had recommended only a triple-orbit mission, allowing Titov to fly five hours in space. However, Korolev pushed forcefully for the extended mission, arguing successfully that anything less would appear too cautious and incremental in character.[66]

A few external factors shaped the timing of the launch. Khrushchev, to the confusion of Korolev and his team, insisted that the launch take place before August 10. Only later did they realize that this date was not arbitrary, but linked to the construction of the Berlin Wall, scheduled to begin on August 13, a secret concealed from the team at Baikonur. At this juncture, Khrushchev had become an avid supporter of the space program and looked upon each successful mission as a propaganda opportunity. Space triumphs were now integral to Soviet foreign policy, a metaphor for the presumed superiority of a society shaped by Marxist-Leninist ideology. Given its strategic importance, the Soviet space program benefited from the largess of the state: Khrushchev had approved a massive outlay of funding for the Baikonur complex, design bureaus engaged in space-related research, and cosmonaut training—the entire costly infrastructure associated with space exploration.[67]

As with most men selected for the cosmonaut corps, Titov was young; he had entered the cosmonaut training at the age of 24. He differed dramatically from Gagarin, being more of an extrovert and outspoken on many issues. He often was restive with the rigors and demands of cosmonaut training. He was more sophisticated than most pilots, possessing wide intellectual interests beyond the narrow sphere of aerospace. He took a keen interest in the arts, expressed great admiration for the American writer Ernest Hemingway, enjoyed quoting long passages of poetry from memory, and displayed an inclination (as circumstances permitted) to discuss political issues. (In the 1990s, he became a member of the Russian Duma.) A native of the Altai region, Titov had an ethnic Russian background. Regarded by his peers as a gifted pilot, Titov's military career, for the most part, had been conventional in nature: He graduated from the Kacha Higher Air Force School in 1957 and then entered active duty as a military pilot in the Leningrad area.[68]

As the pilot of the second Vostok mission, Titov set a new benchmark in manned space travel. His time in orbit, however, did not unfold without challenges. He lifted off from Baikonur at 9:00 A.M. on August 6, 1961. The Vostok 2 booster performed smoothly, inserting the spacecraft into orbit without mishap. Yet no sooner had Titov entered his orbital path than he felt profoundly ill and disoriented. This sensation turned acute, and he felt as if he were flying upside down. One disturbing symptom was the inability to read his instruments or to focus his eyes on the Earth below. To regain a sense of normalcy, he attempted to move about in the tight compartment of the capsule, but this did not improve his condition; if anything, the sudden illness only deepened in severity. Titov had become a victim of spatial disorientation, a consequence of being in near-zero gravity.[69]

By the second orbit, Titov pondered whether he should scrub the mission, asking mission control for permission to return to Earth. Doctors on the ground were fully aware of Titov's situation as they monitored the various sensors recording the cosmonaut's vital signs. During the third orbit, they anxiously inquired about his condition. Titov reported back that he was fine; he was intent on riding out the spasm of disorientation. By the sixth orbit, he attempted to eat a meal from plastic tubes, a menu of soup puree, liver pate, and black currant jam. Still gripped by vertigo and nausea, Titov ate only sparingly, and at one point he vomited. His condition improved with each orbit, and finally he was able to catch some sleep. By the twelfth orbit, he felt nearly normal.[70]

Titov's persistent space sickness had disrupted the planned schedule of experiments, but he was able to resume his work once his health improved. One experiment called for the filming of Earth's horizon for 10 minutes as the spacecraft entered and exited from Earth's shadow. This perspective on the planet was unique, and later images would be published by the Soviets to demonstrate the extraordinary achievement of the Vostok 2 mission. For those on the ground, two internal TV cameras on board showed Titov inside the capsule. The extended orbital mission was largely without mishap, except for a temporary drop in the internal temperature of the capsule, a result of the cooling fans being accidentally turned on at launch.[71]

Reentry, the most perilous phase, offered some further challenges. At retrofire, the Vostok 2 slowed and then began its descent into the upper atmosphere. Shortly after this critical stage occurred, Titov heard a sudden noise, which indicated that the two compartments of the spacecraft had separated on schedule. He became alarmed, though, when he heard a persistent rapping sound, a telltale indicator that the separated instrument compartment was still attached. Both modules entered the atmosphere together, with the instrument compartment eventually burning away. Titov managed to make his homeward-bound descent without difficulty, notwithstanding the initially buffeting that came with the still attached instrument module. The cosmonaut parachuted to safety just past 10 A.M. on August 7. He landed near the small village of Krasniy Kut near the city of Saratov.[72]

Titov orbited the Earth for a total of one day, one hour, and 11 minutes, an astounding record for the time. Subsequent Soviet news coverage offered few details on his discomfort in space, stressing instead the monumental nature of his achievement. In subsequent weeks and months, Gagarin and Titov—the two heroes of the Soviet space program—made a world tour. As the year 1961 ended, the dominance of the Soviet Union in space travel was universally acknowledged.

"FORD VERSUS CHEVY"

As automobile enthusiasts well know, there is no mistaking a Ford for a Chevy. In the space age, there would be no mistaking an Apollo for a Soyuz, either: American and Soviet space rockets and spacecraft mirrored separate design philosophies. All space-age designs required innovative ways to meld form and function. In the end, each tradition had its own style when it came to design.

Livery. Rockets can be painted any color or no color at all, so rocket liveries often reveal something about the priorities and values of their builders. The primary color of Soviet space rockets reflected their very close relationship with the Soviet military missile programs: They were generally painted army green. White accents and bare-metal areas appear variously on later Soviet rockets without strict scheme, perhaps reflecting the somewhat improvisational nature of Soviet engineering.

Rockets developed under U.S. military programs, such as the Atlas and Titan series, were typically silvery bare-metal vehicles because hull paint was nonessential. Black bars and white backgrounds were applied where high-contrast visual tracking marks were needed, but overall these rockets were unpainted. These silver rockets expressed a military attitude of utilitarianism.

The American rockets that made the strongest impression in the public media, however, were white with black markings. This livery defined the popular American image of what a rocket looked like. From the Mercury-Redstone of Alan Shepard to the Saturn IB of Wally Schirra and the great Saturn V rockets of Apollo 8 and all the moon landings, the look was consistent. These rockets all carried black and white livery because they were developed under the supervision of Wernher von Braun's rocket team, which had been using it on their rockets ever since the first successful A-4 (V-2) had lifted off at Peenemünde in 1942. The scheme had originated at the direction of von Braun's mentor Hermann Oberth, but was created for the science fiction rocket seen in Fritz Lang's 1929 science-fiction film *The Girl in the Moon.* The German engineers sometimes justified the high-contrast livery on their rockets

The massive Saturn V first stage in the Vehicle Assembly Building at Cape Kennedy.

as an aid to visual tracking, but the truth was that Wernher von Braun just liked it that way.

Silhouette. When Wernher von Braun envisioned the super-rocket of the future for the book *Across the Space Frontier* in 1952, he conceived a vehicle with a pointed nose like his V-2, but with a wide flaring base that gave the rocket a tapered silhouette like a megaphone.[73] Von Braun tended to avoid counting on "miracle" technology solutions that might never be developed; he wanted to make realistic plans, so he stuck as close as possible to the state of the art. The V-2 had operated on a single engine.

Building a substantially larger rocket engine presented major engineering challenges, so von Braun presumed that the super-rocket would need dozens of rocket motors, including 51 in its first stage alone. This platoon of engines swelled the base of the super-rocket and gave it its characteristic tapered shape.

As the United States entered the space race in earnest, engineers at companies such as Rocketdyne and Aerojet-General *did* produce "miracle technology"; the brilliant development of superior engines greatly exceeded von Braun's conservative estimates. The two-seater Gemini capsule was launched by a Titan II rocket with only two Aerojet LR87 engines at its base. Rocketdyne's ultimate masterwork, the F-1 engine, developed an astonishing 1.5 million pounds of thrust. Even the towering Saturn V moon rocket needed only five such super-engines, not 51. With no need for numerous engines, U.S. rockets could take the simpler form of a cylinder rather than a tapered shape. The straight tube therefore became the basic hull element for most American rockets and gave them their characteristic columnar silhouette, which expresses the advantage of high-power engines—and the advantage of a more powerful economy.

The Soviet Union did not have the luxury of the American super-engines in its advancing rocket development, and therefore generally had to use greater numbers of engines in its rockets to generate the necessary thrust. The R-7 that launched Sputnik bore five engines at its base, and its four strap-on boosters all swelled outward at the base to accommodate them. To provide sufficient thrust for the N-1 moon rocket, the vehicle's first stage required no fewer than 30 NK-15 engines. This mass of 30 engines swelled the diameter of the N-1's base, giving the Soviet rocket the bell-like shape that von Braun had imagined for his book back in 1952. Coordinating the operation of all these engines proved to be an engineering challenge of its own—in four launches, the N-1 never made a successful test flight.

Piloted versus controlled. American and Soviet spacecraft-control procedures also directly reflected basic differences in the two national cultures. American pilots were groomed as extreme incarnations of the American ideal of individualism, and the early astronauts earned reputations as free-spirited and independent. The Mercury 7 astronauts greatly resented the idea that they would lie in their space capsules as passive occupants little more significant than the chimpanzees NASA launched before them. U.S. spacecraft reflected the philosophy of individualism, in that they were under the control of the pilots.

Soviet cosmonauts were likewise held up by the state as icons of the national character ideal. Soviet spacecraft were controlled like Soviet society, by the party, with the pilot normally taking control only in the event of an unanticipated emergency. Yuri Gagarin launched into space on his pioneering flight with the control systems of his Vostok spacecraft locked to prevent him from taking manual control.

Splashdown versus landing. A space-craft plunged into water tends to enjoy the benefit of a much softer landing than one that must hit the ground. Water landings consequently appeal to engineers and space travelers alike, but most of the world's seas are legally "international waters." Early spacecraft could not make pinpoint returns, so a nation had therefore to mobilize a considerable naval force to ensure that its returning spacecraft—especially during the Cold War—was not plucked out of the water by some-one else. In view of the high expense required by ocean recovery, land-recovery methods such as a paraglider parachute and ski-landing gear for the Gemini capsule were studied seri-ously. In the end, however, the United States accepted the ocean-recovery cost and picked up its astronauts at sea throughout the moon race.

The Soviet Union opted for land recovery of its cosmonauts to keep them within Soviet territory. Since their capsules were not designed to protect them from the landing impact, early cosmonauts had to eject at the proper altitude and return to Earth by personal parachute. The interna-tional flight records maintained by the FAI (Fédération Aéronautique Internationale) required pilots to stay with their craft through the entire flight in order for achievements to be registered as official. The space race ran on the prestige of firsts and records, and the Soviet solution to their problematical "parachute gap" was simply to fudge the reports and hide the truth. The official news agency reported for years that Gagarin and his successors had all returned to Earth inside their spacecraft.

Most of the differences between Soviet and American space hardware and operations represent alterna-tive approaches with similar results. However, their different approaches have sometimes led to significantly dif-ferent results. American super-engines helped put astronauts on the moon, but the high costs of that approach brought a quick and total end to the Apollo-Saturn programs. The Soviets tried to reach the moon without fund-ing the development of super-engines for the N-1, and failed. But their eco-nomical solutions had their virtues: America discarded the Apollo space-craft system as too expensive, later discontinued the costly shuttle orbiter that followed it, and is now undertak-ing the costly reengineering of a moon rocket and a crew spacecraft. Through all this, rather like a Volkswagen across the decades, the basic Soyuz spacecraft has remained in production. ∎

PORTALS TO THE COSMOS

"In the evenings, all of us poured out on to the street, away from the light sources, and awaited the appearance of the quickly moving, pale asterisk that we had thrown into the sky," reported Svyatislav Lavrov, an engineer who collaborated with Sergei Korolev in developing the Soviet Union's first generation of missiles. In the 1950s, Lavrov belonged to a small elite of Russians with an insider's knowledge of the Soviet space program, and he expressed his profound sense of pride in the Sputnik triumph. For Lavrov—and a vast global audience in October 1957—the "pale asterisk" in the nighttime sky indeed heralded the new space age.[1]

Sputnik was visible to a global audience, a sudden and awe-inspiring technological feat. Yet the earthly origins of the satellite remained shrouded in mystery, even to most Russians. As late as the Gagarin mission in April 1961, Moscow did little to dispel the mystery, making only vague references to an unnamed "cosmodrome" as the launch site for its space spectaculars. Gagarin kept the launch site secret from his family, and his

OPPOSITE: *Alan Shepard lifts off from Cape Canaveral on the first U.S. manned spaceflight, May 1961.*

mother only learned of his epic flight by listening to the radio.[2] In time, however, the world learned that the Soviets were launching their robotic and manned space missions from a remote spot in Kazakhstan later to be called Baikonur. It was here that Sputnik had been "thrown into the sky," to use Lavrov's words.

The building of Baikonur—as with NASA's parallel facility at Cape Canaveral—gave architectural expression to the new space age as it emerged in the late 1950s. Both sites embodied in scope and purpose a conception more ambitious than a conventional missile test range. Both Baikonur and Cape Canaveral, then and now, are best understood as "spaceports," facilities for the conquest of space. As dueling focuses for space exploration, they were very costly and required a vast work force. Each complex occupied a huge terrestrial footprint with numerous launch pads, oversized rocket-assembly buildings, flight-control centers, fuel plants and depots, training facilities, and housing compounds. Each space center, in turn, necessitated allied institutions to sustain the space exploration program—design bureaus, research centers, manufacturing plants, and a global network of tracking and communications posts. Baikonur and Cape Canaveral—not Sputnik—became the enduring symbols of the new space age.

As portals to the cosmos, Baikonur and Cape Canaveral differed in their public accessibility and their overall style, a contrast that reflected in part two distinct political cultures. Baikonur, simply put, was a sealed-off world, always detached and inaccessible, a military precinct carefully hidden from public scrutiny. Ironically, the cosmodrome was built on the eve of the 20th Party Congress, where so many of the secrets of the Stalinist era were revealed and denounced. This so-called "thaw" under Nikita Khrushchev did not, however, reverse the Soviet practice of maintaining strict secrecy about the emerging space program, given all its organic ties to the military. Those who labored at Baikonur, most notably Sergei Korolev, remained anonymous, heroic figures, for certain, but unknown by name to the Soviet populace.

By contrast, Cape Canaveral—even with its classified military space and weapons programs—operated for the most part in the public sphere, with NASA overseeing an ambitious space program of satellite launches and manned space missions. At the Cape, NASA deliberately showcased the American space program, which was linked to the institutional goal of building public—and critical congressional—support. The American print and television media rewarded NASA's openness with favorable coverage for the most part, routinely casting the astronauts as heroes. The American space program nevertheless remained vulnerable to public criticism whenever well-planned missions misfired. Moreover, from the outset, NASA faced critics who argued that the agency's program was too extravagant and the manned space missions were not cost-effective.

By contrast, the Soviets kept operations at Baikonur under wraps, selecting only certain space exploits for coverage and then through the narrow conduit of official propaganda. Life at Baikonur would also be punctuated by technical mishaps, mission failures, and even the loss of human life. Yet these reversals were routinely downplayed or simply ignored in official Soviet pronouncements.[3]

WHERE THE LAUNCH PAD BEGAN

Early space pioneers had to confront the inherent dangers that came with liquid-fueled rockets. These experimental rockets were never predictable, given their volatile fuel and complex array of motors and pumps. Overheating became a particularly difficult challenge. In the absence of an effective cooling system, a liquid-fueled rocket would explode violently. To fire such volatile missiles required ample space and strict safety measures; these requirements would persist into modern times.

Knowing these realities, Robert Goddard first experimented with a liquid-propellant rocket on his aunt Effie Ward's farm outside Worcester, Massachusetts, in 1926. Later, Goddard moved his laboratory to New Mexico, which offered an expansive landscape for his experiments. The move was made possible through the active support of Charles Lindbergh and the patronage of Daniel Guggenheim, who subsidized Goddard with a yearly grant of 25,000 dollars. Goddard set up his own test range in the desolate Southwest near Roswell. He used his grant funds to pay his salary and the salaries for his assistants and to purchase supplies such as aluminum, steel tubing, liquid oxygen, and other components to build a rocket. Twenty miles into the desert, Goddard built a test stand and rocket-launching tower. Lindbergh described the pioneering rocket facility: "That tower, made out of the galvanized-iron framework of a windmill and pointing skyward from its concrete corner blocks, seemed to me like a huge cannon aimed at an unwary moon. Someday, I felt, a rocket launched from Earth would strike the distant satellite."[4]

The remote desert set the stage for the launch of Goddard's 15-foot-long rocket, dubbed "Little Nell," which had to be hauled to the test stand on a specially constructed trailer. Lindbergh recalled the primitive setting: "We would evacuate tarantulas and orange-spotted black widows from the nearby dugout observation post. . . . An assistant would pour sub-zero liquid oxygen from a big thermos bottle into one of the rocket's fuel tanks. Stations would be taken—one man in the dugout, the rest of us at the control post about a thousand feet away. At a word from the Professor [Goddard] there would be a flame, a roar, and . . . a slender object streaking skyward."[5] With his layout Goddard anticipated the essential architecture for future spaceports, where massive liquid-fueled rockets would launch artificial satellites and deep-space probes and send humans to the moon.

German rocket pioneers followed a parallel path to Goddard in the interwar years, also seeking an isolated locale to launch their rockets. These experiments by Wernher von Braun, Hermann Oberth, Klaus Riedel, and other early German rocketeers were initially conducted at a relatively small test range outside Berlin. Once the group received the active support of the German army, they sought out a larger and more secure test range. By 1937, a secret rocket research center took shape at Peenemünde on an isolated stretch of territory on the Baltic Sea coast. The new facility offered adequate space to perfect the A-4 (V-2) rocket that quickly transformed modern rocketry and set the stage for the space age. The builders of Peenemünde—backed by the financial largess of the German army—constructed a complex that was genuinely functional— an ensemble of administration buildings, laboratories, manufacturing and assembly structures, housing units, warehouses, and concrete test stands. Here rockets could be safely launched across the Baltic Sea for miles. German aerodynamist Peter Wegener remembered the futuristic character of the Peenemünde rocket center when he first encountered it during World War II. Approaching one of the test stands, he had the opportunity to see a launch of a V-2 rocket, which "rose slowly, then bent toward the Baltic to disappear in clouds at great height. I remember an excruciating loud initial bang of the rocket engine, a screeching noise, clouds of deflected exhaust, into which a deluge of water was shot, and a reddish jet issuing from the back."[6] Wegener had observed the world's first fully operational rocket launch facility, a template for Baikonur and Cape Canaveral in later decades.

THE SOVIETS BUILD A SPACEPORT

The concept for the Baikonur cosmodrome was futuristic and ambitious. Once built, the new facility would be the largest in the world. The times were favorable for such a Herculean endeavor, first for the formative military experiments with missiles and then the active program of space exploration. The U.S.S.R. Council of Ministers signed the decree to build the new launch range on February 12, 1955.[7] Within three months, construction work began on the site in distant Kazakhstan, in Soviet-controlled Central Asia. The precise region selected for the test range was located 200 miles southwest of the town of Baikonur. The Soviets deliberately associated the name Baikonur with the new space center as a way to confuse American intelligence. The actual site, the Tyuratam railroad stop just north of the Syr Darya River, was a desolate region—1,300 miles southeast of Moscow, approximately 100 miles east of the Aral Sea, and 500 miles west of Tashkent, Uzbekistan. American U-2 spy plane pilots first located Baikonur by following railroad tracks until they came across the rocket complex during its construction.[8]

Baikonur became a large and complex world during the course of the next three decades. The population may have soared to more than 100,000 by the end of the 1960s, although precise figures remain elusive. The enormity of the cosmodrome can be measured by its array of 52 launch sites, numerous laboratories, factories, schools, palaces of culture, transplanted trees, and artificial lakes. In her book *Kosmos: A Portrait of the Russian Space Age*, Svetlana Boym has described the huge facility as a "Soviet fairy tale come true," with the appearance of a "garden city" in the middle of a vast desert, the locale for "mythic" launches of satellites and cosmonauts into space. Even if concealed behind a wall of secrecy, the legendary spaceport represented a monumental triumph of Soviet technology—very much in the tradition of the great state-run experiments of the 1930s.[9]

Baikonur was part of an entire shadowy world of Soviet technical and industrial centers, or "closed cities," that remained off-limits to outsiders. During the Cold War, the Soviet regime also established ten "nuclear cities" where scientists and technicians worked on warhead designs and produced highly enriched uranium and plutonium for nuclear weapons.[10] Not all Soviet projects that were keyed to industrial or scientific progress were classified, as evident in the aforementioned 1930s Five Year Plans: the construction of the tractor works at Kharkov, the automobile factories at Moscow and Gorky, the great dam on the Dnieper River, the vast steel furnaces in the Don Basin, new industrial cities such as Magnitogorsk, and the building of the Moscow subway.

Walter Duranty, then a newspaper reporter for the *New York Times,* gave an uncritical endorsement to such massive projects, by describing them as the "New Russia" and a "monument to past struggles and to future hopes, a proof of their mighty Today and limitless Tomorrow."[11] Duranty's selective reportage, echoing as it did official Soviet propaganda, ignored the vast human suffering of the Stalin era. The leadership of the Soviet Union did not hesitate to mobilize vast material and human resources in the cause of modernization. The West faced the formidable challenge in the Soviet period of gaining accurate data on key economic and technological trends. Baikonur—if clandestine and remote—mirrored this powerful underlying ethos of the Bolshevik regime. "The dialectic of invisibility and conquest was characteristic of the whole Soviet space program," observed Svetlana Boym. This impulse for secrecy was powerful, even deeply rooted in the Soviet political culture, a pervasive factor in all state-run affairs—large and small. As Boym observed, "the work of thousands of people who made the space program possible—from the Chief Designer to the cleaning women—remained outside official representation."[12]

There were practical considerations as well. In many ways, the new site was selected for its compatibility with the requirements for testing the R-7 ICBM. The area had to accommodate a 300-day launch program per year; such a site could be isolated, but the weather—no matter the extremes—had to permit this robust program of launches. Size was another key factor: The test range should be vast, to allow for the dropping and recovery of rocket stages. Radio guidance for the R-7 required two stations separated by some 300 miles from the launch pad. Rail links, at least initially, were vital to allow the shipment of materials and personnel. Once completed, the new facility would have an airfield and a network of connecting roads and highways.[13]

Baikonur offered a secure and optimal location to catapult rockets and their payloads into orbit. The vast desert region around the rail stop was indeed remote and hidden from any major population center. As it soon became known, rockets could be launched, tracked, and recovered without hindrance. The southern latitude, as far south as any of the other Soviet states, allowed planners to take some advantage from the added velocity of being closer to Earth's Equator, where Earth rotates west-east the fastest.[14] Baikonur replaced the older Soviet test range at Kapustin Yar (near present-day Volgagrad), where the V-2 rocket technology had been tested, along with the R-5 missile. Baikonur would assume historical significance as the locale for the Sputnik launches, the lunar probes, and the early Soviet manned space missions. Today, the fabled cosmodrome remains at the epicenter of the modern Russian space program—the home for latter-day Proton, Tsyklon, and Zenit space launch vehicles.

In April 1955, Colonel Engineer (later General) Georgiy M. Shubnikov assumed overall leadership of the project. Shubnikov had served as the chief of the 130th Directorate for Engineering in the Ministry of Defense. His task was a stern one, given the enormous administrative burdens, and he worked assiduously to meet all the exacting deadlines. The entire project was gargantuan in scale, requiring an intense schedule of work and constant attention to quality control. He directed a steady stream of building materials, engineers, and workers to the site. Shubnikov was assisted by the talented and hardworking Colonel Ilya M. Gurevich, who would spend no less than 20 years of his life in this assignment, only retiring in 1975 owing to declining health. If Korolev held the position of Chief Designer, Gurevich might claim the informal title of "Chief Builder" of Baikonur.[15]

When Shubnikov and his party arrived at Tyuratam, they encountered a forlorn scene: The obscure rail stop consisted of merely a water tower for steam engines, several cottages for railroad personnel, and a cluster of mud huts occupied by local Kazakhs. On all sides of this small oasis, endless desert stretched to the horizon.[16]

Anyone assigned to the Baikonur construction site faced extremes in weather. Aleksei Leonov, the first Soviet cosmonaut to walk in space, remembered vividly the brutal swings of the desert climate. Each new season offered its own measure of punishment. "Hurricane-strength winds reduced temperatures to below −40°F in winter," he recalled. "Once the snow melted, the incessant high winds hurled sand against the buildings with such force that, although we jammed towels against doors and windows, fine dust and sand got in everywhere: in our clothes, eyes, and food." At the height of summer, Leonov told of temperatures soaring to the range of 104-122°F, which forced him and his fellow cosmonauts to wrap themselves in wet sheets as a means to lower body temperature. This remedy, however, had a grim outcome: The moisture attracted insects into their shelter. Scorpions, snakes, and poisonous spiders also abounded. Leonov recalled how death could lurk everywhere in the harsh environs of Baikonur: "I once witnessed a technician, a young captain, being bitten on the neck by a spider. He collapsed and died within minutes. There was nothing we could do."[17] Leonov's experiences echoed the ordeal Goddard had endured in the deserts of New Mexico and, to a lesser extent, those faced by military and NASA workers at Cape Canaveral in the 1950s.

Baikonur evolved only slowly to the point where it could offer any creature comforts to its resident workers. In its untouched aspect, the desert region possessed a certain beauty: "In early spring," Chertok observed, "tiny yellow tulips on the thick clay crust covering the surface were a delight to the eyes." Travel by trucks was possible, but the resulting damage to the engines and undercarriages of these vehicles was often severe. Also, the trucks created deep ruts over the landscape. The first waves of workers to reach the site were housed in prefabricated barracks and wooden houses. The work was exhausting and the living conditions remained primitive during the first two years of incessant work.[18] In the early days, soldiers often were forced to live in dugouts and tents. The construction of a wooden cinema hall and creation of a park added a much-welcomed civilizing touch.

Slowly a town took shape along the Syr Darya River, named *Desyataya ploshchadka* ("Site Number 10"). The town plan called for apartments, a data processing center, administrative buildings, housing units, a department store, and other stores for essential commodities, a hospital, a heating plant, and a power plant. Water in movable tanks was shipped into the settlement until a proper water system from the nearby Syr Darya River had been constructed. A grid of streets and avenues, each lined with newly planted trees, gave Site Number 10 the look of a real town. Residents were housed in typical Soviet-style apartment blocks, erected in five- to seven-floor buildings. Schools appeared once the town was prepared to house families rather than brigades of soldiers and workers. To supply the daunting tasks of construction, a cement plant was set up next to this residential precinct. Later the town would be renamed Leninsk in Soviet times and then, after the fall of communism, Boris Yeltsin renamed the space city Baikonur. By the 1980s, the town boasted a population of more than 120,000 residents.[19]

The initial launch site for the R-7 rocket began to take shape in September 1955. The pad acquired the name "stadium," and it became the locale for a series of historic launches, including the Sputnik 1 artificial satellite and the first manned orbital mission by Yuri Gagarin. The construction of this pivotal launch pad required the pouring of more than a million cubic yards of concrete. A control bunker was built 218 yards from the pad, a secure bastion for rocket technicians that was reputed to be properly concrete-encased to withstand a nuclear attack. This site was later renamed the Gagarin Complex.[20]

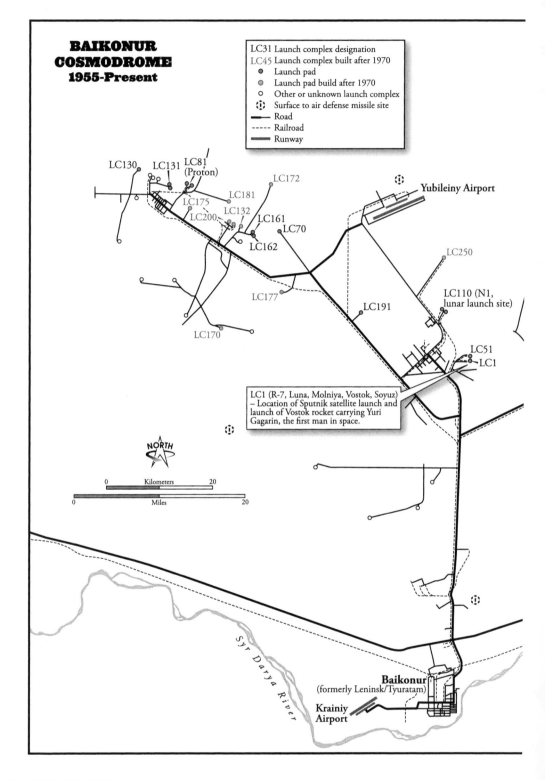

BAIKONUR COSMODROME 1955-Present

LC31 Launch complex designation
LC45 Launch complex built after 1970
● Launch pad
● Launch pad build after 1970
○ Other or unknown launch complex
⊕ Surface to air defense missile site
—— Road
----- Railroad
▬▬ Runway

LC130 LC131 LC81 (Proton)
LC172
LC181
LC175 LC132
LC200 LC161
LC162 LC70

Yubileiny Airport

LC250

LC110 (N1, lunar launch site)

LC177 LC191

LC170

LC51
LC1

LC1 (R-7, Luna, Molniya, Vostok, Soyuz) – Location of Sputnik satellite launch and launch of Vostok rocket carrying Yuri Gagarin, the first man in space.

NORTH

0   Kilometers   20
0   Miles   20

Syr Darya River

Baikonur
(formerly Leninsk/Tyuratam)

Krainiy Airport

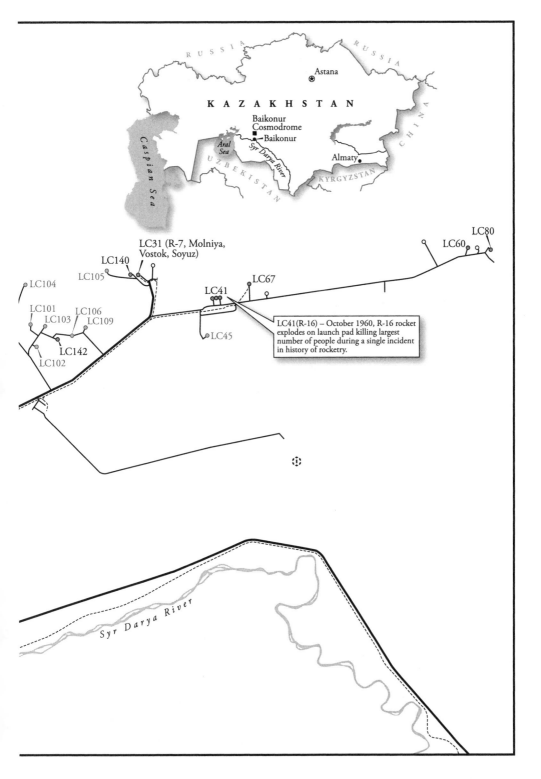

LC80

LC60

LC31 (R-7, Molniya,
Vostok, Soyuz)

LC140

LC105

LC104

LC67

LC41

LC101

LC106

LC103

LC109

LC45

LC142

LC102

LC41(R-16) – October 1960, R-16 rocket
explodes on launch pad killing largest
number of people during a single incident
in history of rocketry.

Syr Darya River

In 1958, a second launch site, named Site 31, was set up for testing of the R-7 missile.[21] Both launch pads were used for space missions in the 1960s. Other launch pads were developed over time, some massive enough to accommodate the Soviets' largest rockets. Baikonur remained an operational missile base, where the latest ICBM designs were tested, beginning with the R-16 in the late 1950s. Over time three launch complexes (left, center, and right in common parlance) were allocated for the use of the major designers: Sergei Korolev, Vladimir Chelomei, and Mikhail Yangel. The center complex was devoted to projects from Korolev's OKB-1 group, a sequence of remarkable and ambitious projects that extended from the R-7 through the Vostok to the N-1 rocket intended to take cosmonauts to the moon. The Soviets built a massive assembly building at Baikonur that rivaled in size the Vertical Assembly Building at Cape Canaveral, where America's massive Saturn V moon rockets were assembled. Linked to Baikonur was a network of control and tracking stations, all run by the Soviet military and positioned at strategic points across the vast continental expanse of the Soviet Union.

The United States suspected that the Soviets had built the massive Baikonur facility, but it took time and effort to get a clear focus on the new facility. First, American intelligence learned of the construction of the top-secret facility in the years 1955-1956, and then took steps to photograph it. The only maps available to the CIA at the time were old German military maps fashioned from aerial reconnaissance in World War II. Hence, the U-2 aerial photography was deemed essential and it was pursued in a systematic fashion. By the summer of 1957, U-2 spy planes had taken the first high-altitude photos of the emerging cosmodrome. This concerted effort at aerial photography was not without risk—or consequences for the tense stalemate between the United States and the Soviet Union in the Cold War. Francis Gary Powers, flying a U-2 spy plane over Soviet territory in May 1960, was shot down during a mission to capture images of the Baikonur complex. Later, reconnaissance satellites offered American photograph interpreters more detailed data on space activities at Baikonur. In December 1967, an American reconnaissance satellite photographed the construction of Complex J, the new spaceport for the massive N-1 rocket to carry Russian cosmonauts to the moon.[22]

Life at Baikonur was punctuated with more than one tragic disaster. On October 24, 1960, the experimental R-16 ICBM, a rocket designed by Mikhail Yangel, exploded on the launch pad at the cosmodrome. Several hundred soldiers and technicians had gathered at the launch pad for the occasion. Marshall Nedelin, the head of the Soviet Union's Strategic Rocket Forces, decided to observe the launch close to the rocket, notwithstanding the warnings he received concerning the inherent dangers of being in such proximity to an untested rocket. Nedelin dismissed such advice, reportedly saying, "What's there to be afraid of? Am I not an officer?" His bravado led to his death. The power and intense heat of the explosion obliterated most of the victims. Marshal Nedelin's remains were identified only because he had worn a distinctive badge. The exact number of casualties in this catastrophic explosion would be a matter of speculation, with estimates of about 130 personnel lost.[23]

A second tragedy befell the Soviet space program in March 1961, not in Baikonur, but in Moscow at the Institute of Aviation and Space Medicine. Cosmonaut trainee Valentin Bondarenko was burned to death in an accidental fire in an isolation chamber. His untimely end cast a pall over the manned space program at Baikonur.[24]

The Baikonur cosmodrome continued to propel the Soviet space program forward in the 1960s with its Vostok series. Vostok 3 and Vostok 4 executed a dramatic rendezvous in orbit. From Baikonur, the first woman, Valentina Tereshkova, flew in space in June 1963. In 1964, the Soviets launched three cosmonauts aloft into orbit in the Voskhod mission, another dramatic and high-risk endeavor to remain ahead of the curve on "space firsts." On the Voskhod 2 mission, Leonov became the first human to "walk" in space. By 1965, just months before his death, Korolev had launched the new Soyuz series to experiment further with rendezvous and docking techniques. The ill-fated work on the N-1, the Soviet parallel lunar rocket to the Saturn V, occupied center stage at Baikonur in the late 1960s, even as the NASA Apollo program gained momentum in the race to be the first to land humans on the moon.

The extraordinary Baikonur facility remained at the epicenter of the Russian space program as it evolved in the post-Apollo context. The Apollo-Soyuz mission in 1975, the Mir space station, and the one-time pilotless launch of the Buran (similar in design to the American space shuttle) in 1988 constituted important milestones that defined the capabilities of the Russian space program. Hard times visited Baikonur in the wake of the fall of the Communist regime in 1991. By 1995, funding dropped to just 10 percent of the 1989 level. Conditions at Baikonur worsened with a steep decline in the standard of living. A major riot occurred in February 1992. In time, though, conditions improved. By the turn of 21st century, Baikonur had regained some of its former stature. No less important, the legendary cosmodrome ceased to be a sealed-off domain. Westerners now routinely tour Baikonur, even participating in joint ventures—a sharp contrast to the pioneering era of the 1960s.[25]

FIRST DAYS AT CAPE CANAVERAL

In the immediate aftermath of World War II, the Army's White Sands test range had been an ideal setting for the American rocket program to mature. As time passed, however, there was increasing interest in alternative test facilities for both civilian and military rocketry. Moreover, the program at White Sands proved to be limited in nature. The city of El Paso, Texas, and the United States–Mexico border lay little more than 50 miles south of the test site; moreover, given the range of rockets, the test site was miniscule, being little more than 100 miles long, north to south.

The perils associated with a small test range became evident on the evening of May 29, 1947, when a V-2 lifted off from the southern part of the range on a planned northbound trajectory. Soon after liftoff, the rocket veered off course, turning south toward Mexico. The abrupt change in direction was caused by a malfunctioning gyroscope in the rocket's guidance system. Now wildly off course, the V-2 flew over El Paso, crossed the Mexican border, and crashed just outside a cemetery near the city of Juarez. The time of the ill-fated launch coincided with a fiesta in the Mexican city, but the explosion—leaving a crater 30 feet deep and 50 feet wide—did not kill or injure any of the local inhabitants. The United States quickly sent formal apologies to the Mexican government. Officials, including Army Chief of Staff Dwight D. Eisenhower and Secretary of State George Marshall, called White Sands to express concern. When American technicians reached the crash site, though, they discovered that entrepreneurial Mexicans had already set up a souvenir stand to sell "authentic" missile fragments to tourists.[26]

Despite these mishaps, from its creation in late 1945, White Sands was a vital and productive center for the Army's missile program. Wernher von Braun and his team and others made substantial gains in the development of powerful rockets and missiles, a research program that grew out of the rocket technology first developed by Germany in World War II. The effort served not only military but scientific ends as well. From White Sands, a series of sounding rockets probed the upper atmosphere. These pioneering launches allowed for a new scientific understanding of near space. Moreover, the resulting spectacular photography of the Earth clearly showed its curvature and topographical features.

In the fall of 1946, the Pentagon began a search for a new test range, to be called the Joint Long Range Proving Ground. The quest for a new test site was justified as necessary to accommodate the range of a whole new generation of missiles. The underlying purpose was to build an effective ICBM, and such a missile required an expansive arena for testing. Several sites were considered seriously—in locales as diverse as Washington State and Florida. Even as the site selection committee prepared to make its final choice, the V-2 from White Sands made its ill-fated crash in Mexico. That incident led to elimination of a California launch site, which would have placed missile testing too close to Mexico's Baja California peninsula, a prospect unacceptable to the Mexican government. The proposed Washington State site was rejected due to its poor weather and isolation. Other candidate sites also fell by the wayside for various reasons, setting the stage for the selection of a 15,000-acre tract of land on Cape Canaveral, Florida, located midway between Jacksonville and Miami. This site was part of a barrier island off Florida's Atlantic coast. The Cape Canaveral site was isolated and remote, except for an abandoned naval air station and an old lighthouse constructed in 1848. Spanish nobleman Ponce de Leon is credited by historians as having been the first European to explore the area—then occupied by Native Americans—in 1513. The region was characterized as rustic and barren, overgrown with thick grass and sharp-edged palmetto scrub along with miles and miles of sand. In the absence of human habitation, the region was populated by numerous scorpions, ants, alligators, snakes, horseflies, and intense clouds of mosquitoes.[27]

Cape Canaveral had attributes that contributed mightily to its selection. As real-estate agents often note, the three most important factors to consider when looking at property are, "location, location, and location." In at least three essential ways, Cape Canaveral offered just that. First, it was ideally placed to take advantage of the Earth's rotational speed. By firing a satellite-carrying missile eastward—that is, in the same direction that Earth is rotating—the rocket gained a head start on achieving orbital speed. Second, Cape Canaveral was relatively close to the Equator, where the planet's rotational speed is greatest. Third, once a rocket was fired out over the Atlantic, headed generally southeast, nothing—save a handful of small islands, mainly in the Caribbean—stood in its way for 5,000 miles. Those islands, stretching from the Grand Bahamas to Ascension Island in the South Atlantic, made perfect locations for tracking stations to follow the missiles' course. The U.S. government negotiated with Great Britain for rights to place the tracking stations on the Grand Bahamas, Grand Turk Island, and Ascension Island. In addition, Cape Canaveral was essentially flat, making runway construction relatively simple. On von Braun's first visit, the Cape reminded him of his former missile test site at Peenemünde, on the Baltic Sea. Both sites, he noted, were isolated and inaccessible.[28]

BUILDING A WORLD-CLASS TEST RANGE

President Truman signed legislation formally creating the Joint Long Range Proving Ground in May 1949, and then placed the operational control of the test facility under the newly independent U.S. Air Force.

Actual construction began in May 1950. Those who worked on the new test facility would encounter real hardships: Before serious construction could begin, for example, a massive mosquito extermination effort, including aerial spraying, was undertaken. The location was primitive, as a Douglas Aircraft Company chief engineer later recalled of his visit to the planned site of a Thor missile launch pad: "We climbed on a bulldozer. Some old Florida guy drove us out there (on a paved road surrounded by marshes and wild scrub), with water moccasins falling off the blade. He said, 'This is where the pad's gonna be.'"[29] Besides an access road into the Cape, the Army Corps of Engineers, working with private contactors, first constructed a 100-foot-wide concrete launch pad, which was completely surrounded by sand. Known as Pad 3, the primitive support facilities for the Bumper V-2 rocket included a control-center blockhouse, a mere 300 feet from the pad, made from "plywood and protected by walls of sandbags." (By contrast, the White Sands test range blockhouse boasted a 27-foot-thick concrete roof.) The Bumper V-2 was composed of a V-2 topped with a WAC-Corporal research rocket, creating the United States' first two-stage space vehicle.[30]

July 24, 1950, dawned as just another day of unbearable heat, very high humidity, and unceasing attacks on Cape workers from surviving mosquitoes, of which there were still uncountable tens of millions. A Bumper V-2, white with black markings, stood ready on a small launch pedestal on Pad 3 to take its place in history as the first rocket launched from Cape Canaveral. It had taken the place of another Bumper V-2, which had misfired on the launch pad days earlier. Photographs of the initial launch show just how much has changed since Cape Canaveral's early days. Standing next to the Bumper V-2 was a painter's tall metal-tube scaffold pressed into service as the rocket's service tower. Mounted on rollers, it had three wooden platforms near its top that surrounded the rocket during pre-launch preparations, permitting access to the small upper stage. Just before 9:30 a.m. the rocket lifted off flawlessly. Another photograph shows several cameramen filming the event from what appears to be an impossibly close, and totally unprotected, position near the launch pad. Despite its strong start, the flight was a disappointment overall;

the second-stage WAC-Corporal failed to fire.[31] The launch nonetheless marked the start of a unique new enterprise, the American Space City.

In August 1953, the Army began testing its Redstone ballistic missile from a newly constructed Pad 4. Unlike Pad 3 with its jury-rigged painter's scaffold, the Pad 4 complex had a steel service structure—actually a specially modified oil derrick—that was more than twice as tall as the 69-foot Redstone. The missile was delivered to the pad on a huge transporter-erector and raised to a vertical position. The Redstone was then enclosed by a series of covered servicing platforms built into the gantry, which was used by technicians to access the missile at several levels for pre-flight preparations. Before the Redstone was launched from a firing stand, the giant service tower, mounted on rails, was rolled well away from the missile to protect it in case the missile exploded on the pad. Pad 4 supported several early Redstone launches pending completion of Launch Complex (LC) 5 and 6 in 1955-1956.

This two-pad LC was a major step in the evolution of the Cape's facilities, serving as the prototype for the many complexes that were to follow. In 1961, Americans watched Alan Shepard and Virgil Grissom blast off from LC-5 on their suborbital Mercury Redstone flights. The twin pads shared control facilities and offered essential backup in case an explosion rendered one of the pads inoperative. The central blockhouse was heavily fortified to withstand damage from a missile blow-up and had small, thick glass windows affording clear views of both pads. A wide variety of instruments gave technicians vital information on the rocket's various systems and readiness for launch. An essential activity was continuous monitoring of the missile's liquid oxygen supply; if too much of the ultracold liquid oxidizer was vented through a pressure release valve before launch, a tanker truck would have to top off the supply. The service towers were similar to the one installed on Pad 4.[32] A steel umbilical tower next to the launch pad provided electrical power and vital fluids to the missile; the connecting cables fell away at the instant the rocket lifted from its pad.

A typical launch in the late 1950s offers a snapshot of Cape operations in the early years. At that time, one of the most critical missiles being tested was the Atlas ICBM. The military viewed its rapid development and deployment as an essential deterrent to a possible Soviet attack. Befitting its status, no less than four launch pads were erected for Atlas testing, soon to be joined by four adjoining pads for the Titan, America's second ICBM. Collectively, the eight pads, all located close to the Atlantic Ocean, were known as "ICBM Row."[33] They were a most impressive, striking sight, each boasting a huge steel gantry and accompanying support facilities, surrounded by vast expanses of untouched palmetto scrub. Nearby, a growing number of other LCs supported the Jupiter and Thor intermediate-range ballistic missiles (IRBMs), Vanguard, and other missile programs.

The Atlas was build by General Dynamics Corporation's Convair Division plant in San Diego, California. Somewhat incongruously for a weapons system that would fly thousands of miles to a target in 30 minutes, it was hauled across the country in a special tractor-trailer. Air police from the Air Force and a phalanx of Convair technicians accompanied the monster cargo shipment on a 2,700-mile journey, taking more than a week. The Air Force later flew the 75-foot missiles directly to the Cape on a huge cargo plane, using the Cape's landing strip, known as the "Skid Strip."[34]

Upon arrival, the Atlas was delivered to a collection of assembly and checkout hangars maintained by the various manufacturers and contractors working at the test center. Most of these general-purpose facilities, located in the Cape's industrial park area, were built using a standardized design. They typically contained sophisticated testing equipment and cranes to move the missiles around.[35] Once an Atlas was checked out and prepped to the satisfaction of the Convair technicians, it was taken to a launch pad, still on its trailer, and lifted to an upright position on a launch ring in front of its 12-story gantry. The gantry's seven retractable work platforms then embraced the missile, allowing the technicians full access to the huge rocket as they prepared it for launch. In the early days of testing ballistic missiles at the Cape, that task could consume weeks. Only then would the gantry be pulled back on its rails to a position several hundred feet from the pad, leaving the Atlas standing alone. The rocket was then filled with a highly volatile mix of kerosene-based RP-1 rocket fuel and the fuel's oxidizer, liquid oxygen, or LOX, which must be kept at nearly 300°F below zero.[36]

The long countdown preceding launch was conducted from a heavily reinforced blockhouse crammed with instrument-studded consoles, control panels, and other equipment used to monitor the rocket both before and after its launch. The blockhouse crew viewed the launch pad through periscopes. The string of tracking stations extending from the Cape across the Caribbean and well beyond were readied to follow the missile's flight. Typically, after a number of countdown "holds" to address technical problems, the Atlas prepared for launch; its engines at full thrust and straining at the clamps holding it to the pad. Below the launch pad, enormous steel "flame deflectors," each twice the height of a man, caught the exhaust flames from the Atlas's engines and channeled them away from the missile. Water was sprayed over the deflectors at a rate of up to 35,000 gallons per minute to cool off the flaming exhaust, converting it into giant clouds of steam. The clamps were released, and the giant ICBM rose from its pad, its engines producing an incredibly loud roar. In the blockhouse, the launch crew anxiously monitored the Atlas's progress.[37] Unfortunately, in those early days of testing America's long-range ballistic missiles, the flights very often ended shortly after launch with an explosion caused by an internal malfunction or the Range Safety Officer (RSO) detonating an onboard explosive charge. The RSO's job was to closely monitor the missile's flight and destroy it if it strayed from the carefully programmed flight plan. Over time, the Cape's launch record improved greatly.

In contrast to the secrecy at Baikonur, the civilian space program at Cape Canaveral unfolded in full view of the public and the media. Crowds routinely gathered on the nearby beaches to watch the launches. While reporters with cameras were not regularly allowed into the launch site proper until the late 1950s, long camera lenses and binoculars provided significant coverage of launches from the beginning, including some spectacular failures. These included on-pad explosions of fully fueled early Atlases and Thors, as well the memorable failure of the first American attempt to orbit an Earth satellite on a

Vanguard rocket in 1957. More than once, Cape observers watched in horror as a giant missile turned around in flight and headed back toward the ground. One such incident in July 1958 involved a NASA attempt to orbit an Explorer satellite using a Juno II booster based on the Army's Jupiter IRBM. Almost immediately after liftoff, the Juno II deviated sharply to the left, departing from its planned vertical path. The RSO destroyed the missile barely five seconds into the flight. Trailing flames from the detonation of the explosive charge, the missile plunged earthward.[38]

CAPE CANAVERAL'S NEW ROLE AS A SPACEPORT

NASA became a vital part of life at Cape Canaveral in 1958. The Air Force, then the chief occupant of the remote site, provided NASA with the launch pads, rockets, and other facilities it needed to conduct its space missions.[39] That was only fitting, because Air Force rockets played an essential role in the new NASA-run space program. For example, NASA's early Pioneer missions to the moon and beyond were carried into space on Air Force (and occasionally Army) rockets such as the Thor-Able, Delta, and Juno II. When NASA's manned spaceflight program took shape, it relied on versions of the Army's Redstone and the Air Force's Atlas and Titan II ICBMs for Project Mercury and Project Gemini. Project Mercury orbital flights (using the Atlas) and all Project Gemini flights (Titan II) were all launched from the legendary "ICBM Row." However, following President Kennedy's proposal to send Americans to the moon, NASA increasingly required its own separate launch facilities on a scale much larger than what existed at Cape Canaveral.

NASA's arrival would sharply accelerate the enormous economic benefits of space and missile programs for the communities around Cape Canaveral, a cycle that had begun with the military missile programs. As early as mid-summer 1957, *Time* magazine quoted a local resident on the area's prosperity: "'We've all got rocket fever here,' said the manager of the Starlite Motel in Cocoa Beach (near Cape Canaveral). . . . 'Everything centers on the Cape.' . . . It is a land where highways are likely to be blocked so that trailers can haul their menacing, canvas-shrouded packages to the secret precincts beyond the gates." The article noted that the population of Brevard County, home of the Cape, had shot to 70,000 in 1957 from 23,600 just seven years earlier. Land values were up as much as 500 percent, and in towns like Melbourne, Titusville, and Cocoa Beach, there appeared to be no end in sight.[40]

By July 1969, with Apollo 11 en route for the first manned lunar landing, Brevard County's population had grown to an astonishing 250,000, including 35,000 engineers and technicians. But the good times could not last forever. Only a year after Apollo 11, and well before the last Apollo mission in December 1972, NASA administrator Thomas Paine said publicly of his agency, "We are at the peril point." He was referring to sharp reductions in NASA's budget, with repercussions extending well beyond the space agency's hardware and missions. The boomtowns around the Cape were hard hit by the resulting fall in NASA and space contractor spending. Vacant and boarded-up homes, businesses, and offices soon appeared. With thousands of laid-off space engineers and technicians looking for jobs, resumé writing became among the most lucrative opportunities in the county.[41]

But in March 1965 the prosperity had received a significant boost. By this time the Cape had been renamed Cape Kennedy. In November 1963, just days after President Kennedy's assassination, President Johnson ordered that Cape Canaveral (the geographic area) and the NASA launch center be renamed in honor of the slain president. The NASA facility became known as the Kennedy Space Center. The Air Force later renamed its Cape Canaveral launch facilities Cape Kennedy Air Force Station. (However, in 1973, in response to public sentiment in Florida, the original historic name of the cape was restored, though the Kennedy Space Center retains that name. The Air Force launch facilities were renamed Cape Canaveral Air Force Station.) In March 1965, the future of the Cape complex was shaping up on Merritt Island at NASA's Kennedy Space Flight Center, across the Banana River from the Air Force's huge missile complex.

With full responsibility for Project Apollo, NASA set aside some 111,000 acres of land on Merritt Island to construct assembly and

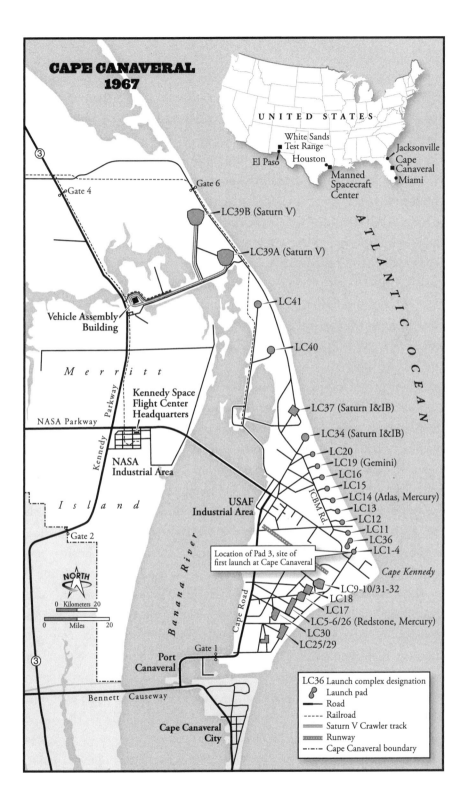

CAPE CANAVERAL
1967

UNITED STATES

White Sands
Test Range
El Paso   Houston   Jacksonville
                      Cape
           Manned   Canaveral
           Spacecraft Miami
           Center

③

Gate 4   Gate 6

                LC39B (Saturn V)

                LC39A (Saturn V)

                    LC41

Vehicle Assembly
Building

                    LC40

M e r r i t t

Kennedy Space   LC37 (Saturn I&IB)
Flight Center
Headquarters   LC34 (Saturn I&IB)
NASA Parkway   LC20
                           LC19 (Gemini)
NASA   LC16
Industrial Area   LC15
                           LC14 (Atlas, Mercury)
I s l a n d   USAF   LC13
                       Industrial LC12
Gate 2   Area   LC11
                                   LC36
NORTH   LC1-4

0 Kilometers 20   Location of Pad 3, site of
                       first launch at Cape Canaveral   Cape Kennedy
0   Miles   20

③   LC9-10/31-32
                                   LC18
                                   LC17
                       Gate 1   LC5-6/26 (Redstone, Mercury)
Port   LC30
Canaveral   LC25/29

Bennett Causeway
                                   LC36 Launch complex designation
                                    ⚲ Launch pad
Cape Canaveral   ——— Road
City   ----- Railroad
                                    ══ Saturn V Crawler track
                                    ▓▓▓ Runway
                                    —·—· Cape Canaveral boundary

ATLANTIC OCEAN

Kennedy Parkway

ICBM Rd.

Banana River

Cape Road

launching facilities for the mighty Saturn V moon rocket. It was going into business for itself, in a rather large way. A 1965 magazine article notes: "Beyond the Banana River, NASA is building new production-line facilities . . . their specifications are so superlative as to strain belief. The 552-foot Vehicle Assembly Building (VAB), for example, will have four rocket bays behind the tallest doors in the world, and more interior volume than the Great Pyramid of Cheops, or the Pentagon and the (Chicago) Merchandise Mart combined." In somewhat breathless terms, the article discusses various aspects of the assembly and movement to the launch pad of the Saturn V, including the "biggest windshield wipers in the world" on the cab of the crawler-transporter that would move the huge rocket. The transporter would carry its massive load on "a special crawlway, as wide as the New Jersey Turnpike and almost 8-feet thick (to support the 17.5-million pound combined weight of the transporter and its load)."[42]

A network of support facilities was linked to Cape Kennedy. NASA administrator Jim Webb created the new Manned Spacecraft Center (MSC) in Houston, Texas. In 1973, this pivotal component in the NASA manned space program was renamed the Lyndon B. Johnson Space Center. The new facility was given responsibility for all human space-flight missions, the training of all astronauts, and the development of the spacecraft needed to carry out the manned missions. Beginning with the second manned Gemini flight, the MSC also functioned as Mission Control for all manned spaceflights, taking over the monitoring and control of these missions once the actual launch took place at the Cape.[43]

The Kennedy Space Center has evolved over the years, as the American space program has taken on new missions and priorities. At the NASA facilities on Merritt Island, the Vehicle Assembly Building complex supports the Space Shuttle, carrying out its final missions to support completion of the International Space Station. NASA also continues to launch a wide variety of unmanned scientific satellites. For these missions, NASA uses launch pads "borrowed" from the Air Force. The Cape Canaveral Air Station pads are also used for commercial space launches of communications and other types of satellites as well as military satellite missions. Compared to the busy Cold War years of high-volume military missile tests from the Cape, many fewer launches take place each year.[44]

Viewed together, the American and Russian space cities represent a stunning triumph: They are portals to the cosmos. They provide separate departure points for both robotic and manned space missions, past and present, which continue to shape our understanding of the universe, even as they have advanced science and technology. These space portals, though evolving from two radically distinct political systems in the context of the Cold War, have become legendary. Yet they bear an underlying similarity in architecture and purpose. Together they symbolize the spirit of human exploration.

THE GREAT ESCAPE

We sometimes think about "space" as equivalent to altitude: If one goes so high, one is in space. In practical rocketry, however, reaching space is usually most concerned with attaining orbit, and this is more reliant on speed than altitude. Some of the experimental V-2 rockets launched at White Sands Missile Range rose up to the edge of space—high enough to photograph the curvature of the Earth—but they fell right back down again. Science fiction aside, one does not start floating as soon as one reaches "space altitude." Even once it has achieved the necessary speed, a vehicle will still fall like the V-2s did, but at orbital velocity it is moving forward so quickly that it "falls" around the curve of the Earth, and continues to do so. This is why astronauts in orbit experience the free-fall sensation of jumping off a diving board the entire time they are in orbit: They are literally falling the whole time. This is also why a rocket leans over sideways so soon after rising off the launch pad; it is aiming into its orbital speed run.

Orbital mechanics primarily have to do, therefore, with achieving very high velocities rather than very high altitudes. Space starts a mere 62 miles up, by international convention. Yuri Gagarin orbited the Earth at an altitude of 188 miles, close to common space-shuttle mission heights, while the Apollo spacecraft used a "parking orbit" on their way to the moon at an altitude of just 115 miles. Orbital velocity for most of these missions is around 17,500 miles per hour.

Faster still than orbital velocity, "escape velocity" is the speed necessary to break free of the Earth's gravitational pull. Whether it is a tiny satellite or a large manned spacecraft, escape velocity will be the same: about 25,000 miles per hour. In parking orbit, Apollo spacecraft used their third-stage engines to increase their speed from orbital velocity to escape velocity, to break away and begin the journey to the moon.

A rifle bullet travels at up to 2,700 mph, a fact that helps us appreciate the force needed to accelerate an entire space vehicle to 17,500 mph. That being the case, one needs all the assistance one can get. Simply choosing the location for a launch site greatly affects how much energy it will take

The Florida peninsula, viewed from Apollo 7, October 1968.

to get a rocket into orbit, because a rocket can "borrow" speed from the rotation of the Earth itself. The Earth rotates toward the east, at a speed that varies depending on latitude. A few yards from the North Pole, the rotation speed is extremely slow, but at the Equator, the planet's surface is moving at approximately 1,038 miles per hour. Rocket engineers are only too happy to take advantage of that fact, so rockets normally launch to the east, to build upon planetary rotational velocity. Rocketeers will also ask for their launch pads to be built as close to the Equator as possible, to gain as much free speed as possible. In 1952 Wernher von Braun reckoned that the rotational assist factor would be so important as to require the United States to site its main launch complex on Johnston Island in the Pacific Ocean.[45] The high costs of transporting everything to this distant point would be offset, he imagined, by the gain in free energy provided by a more equatorial location. The rotational assist principle made novelist Jules Verne in 1865 to site his imaginary U.S. launch complex in Florida, the southernmost continental state. High logistical cost estimates for a Pacific Island site put the real NASA Kennedy Space Center in the state of Verne's prediction, naturally on the Atlantic coast, since the launches would be toward the east.

The Soviets were stuck with a less advantageous geographical situation, not having any location within the U.S.S.R. that was especially southerly. For maximum rotational assist they had to settle for a remote site in Kazakhstan. At 45°57' the Baikonur cosmodrome is at the same latitude as North Dakota. An even more northerly cosmodrome was built at Plesetsk (at the same latitude as Alaska), to serve for satellites launched in polar orbits, which could not use the rotational assist and for Soviet ICBMs. The United States built a corresponding polar orbit and missile launch site on the California coast north of Los Angeles, at Vandenberg Air Force Base.

Physicists in the 19th century were perfectly capable of calculating thrust equations, and in working out the amount of thrust contained in a given quantity of liquid rocket fuel, they thought they had proved that an orbital rocket could never be built. They discovered the paradox that any rocket large enough to hold a given quantity of fuel would be too heavy for that fuel to get it into orbit. Make the rocket bigger or smaller as you like, any real-world construction sturdy enough to contain the fuel would be too heavy for that fuel to drive it to orbital velocity. The physicists and engineers tried many approaches to the problem, but it seemed intractable.

Years later when Wernher von Braun's team developed the V-2 rocket in Germany, they made so many engineering advances that British intelligence dismissed early reports of the rocket as fiction, but it was still constrained by the old paradox equations. While a modified V-2 could reach into space, it could only fall back down again afterwards. But von Braun knew where to go from here.

Russian theorist Konstantin Tsiolkovsky had cracked the paradox early in the 20th century by inventing the idea of *staging*: A rocket built with segmented fuel tanks could drop off each fuel tank as it ran dry, reducing the amount of weight the remaining fuel had to carry. By the last stage, the fuel would be driving only a small vehicle, and *this* had a chance of making it into orbit. The engineering of a sufficiently light-weight rocket vehicle would still be tough, but it would not be impossible. In staged rockets lay the promise of achieving orbit.

This was precisely the solution employed during the space race, which saw the first satellites driven into orbit by multi-staged rockets. The Soviet R-7 consisted of a core rocket with the Sputnik satellite perched in the nose cone; four strap-on additional booster components provided the launch thrust and were dropped when depleted, leaving the center "sustainer" engine to drive the rocket still higher with its lighter load. America's first satellite, Explorer 1, was carried into space by a vertical sequence of four stages, if you count the satellite's own built-in rocket. In the beginning, these rockets needed every last ounce of thrust to succeed.

Placing heavier payloads into orbit would take larger rockets, which both the United States and the U.S.S.R. proceeded to build over the next several years. Engineers on both sides sought creative solutions to increasing the performance of their rocket designs. Just making a rocket strong enough to stand up required a minimum amount of structure, of course— or did it? The American designers of the Atlas rocket had the idea of thinning the tank design to save weight until the hull was no thicker than a dime. Building this design gave them in essence a giant stainless steel balloon. This super-thin tank could not even stand up under its own weight, and von Braun's conservative German engineers were very dubious of the radical design. The trick was that the rocket built this way would stand up just fine as long as it was *pressurized* with air or fuel. Like a balloon, the tank became rigid when filled and pressurized, enough so that it would withstand even launch stresses. The Atlas actually worked and eventually well enough that John Glenn flew in his Mercury capsule atop such a "balloon" rocket.

Engineers have since sought further rocket performance in light-weight fuels, such as liquid hydrogen, and in more powerful rocket engines of many types. In the last 50 years, however, the basic facts of rocket engineering have not changed, and to attain orbit, rockets still rely on multiple stages and all the Earth-rotational launch assist they can get. ∎

THE SPACE RACE QUICKENS

On February 20, 1962, the eyes of the nation focused intently on Cape Canaveral: On this launch date the first American would be catapulted into Earth orbit. Mercury astronaut John H. Glenn, Jr., had received the coveted nod to make the flight. An Ohioan by birth, Glenn had been a Marine pilot, a combat veteran of World War II and Korea, and a celebrated test pilot. In July 1957, Glenn set a new transcontinental speed record of 3 hours, 23 minutes, and 8.4 seconds, a remarkable aerial feat—his average cruising speed was supersonic. NASA chose Glenn for this high-risk space mission in no small measure for his impressive blend of flight experience, technical skill, and ebullient personality.

An Atlas missile—upgraded and "man-rated" for NASA use—had been selected as the booster rocket. Atop the Atlas, Glenn occupied a cramped seat in Friendship 7, the small cone-shaped command module. News crews from the major networks had descended on the Cape in large numbers, offering continuous TV coverage from dawn to dusk.

OPPOSITE: *John Glenn poses with the Friendship 7 capsule he rode into orbit, February 1962.*

The warm temperature, a comfortable 70°F, attracted a huge crowd of onlookers who had camped out on the nearby beaches to observe the launch. Cape Canaveral had acquired new fame as the staging area for America's still-troubled space program. In the four years after Sputnik, NASA had imposed its footprint on the Cape with the construction of assembly buildings, hangars, launch pads, and support facilities.[1] Most Americans viewed the successful launch of Project Mercury's Friendship 7 as a way to recoup national prestige and narrow the perceived space gap with the Russians.

The long-anticipated orbital mission by Glenn had been burdened with no less than 10 postponements. NASA first announced January 16, 1962, as the date for the launch. Faced with technical difficulties with the Atlas fuel tanks, NASA had abandoned the original date. New launch times followed in a dizzying pattern of announcements and last-minute cancellations, seemingly issued day after day. NASA technicians wrestled with an endless round of technical snafus and bad weather. For Glenn and the vast audience of onlookers, frustration reached a crescendo on January 27, when the countdown reached T minus 20 minutes only to be cancelled, this time by an act of God when heavy clouds suddenly swept over the Cape Canaveral complex.

The month of January gave way to February as NASA technicians struggled with a fuel-leak problem on the Atlas booster. Having solved this problem, NASA set February 15 as the launch date, only to abandon the effort with the return of inclement weather. Finally, February 20 arrived full of promise—no apparent technical glitches and excellent weather conditions. The delays—played out in full public view—had reinforced the widespread perception that the United States, indeed, was behind the Russians. The year before, the Russians had made two orbital flights: The first, by Yuri Gagarin, had established a dramatic milestone, the first passage of a human into outer space; the second, by German Titov, had recorded no fewer than 17 orbits of the Earth. Even though overshadowed by the Russians, NASA persisted with its highly scripted plans, offering assurances of ultimate success and seeking to build public confidence.

Glenn brought to the mission the dedication and élan of a test pilot. He appeared to many as a fitting embodiment of the heroic aura surrounding the Mercury astronauts. Glenn knew of the risks associated with the February 20 launch. At the time, he made no reference to this haunting specter, but decades later he confessed to nagging doubts he—and others—felt about the Atlas booster. This general apprehension stemmed from an incident in their first months of training, when the entire company of Mercury astronauts had been flown to the Cape to observe an Atlas missile launch. At the time the Atlas had undergone a series of modifications to adapt it for use as the booster for future Mercury missions. This process of "man-rating" the missile appeared complete, or nearly so.

For the launch of the Atlas, the astronauts were stationed in the camera platform area, a mere 1,200 feet from the launch pad. The night was clear and cloudless, perfect for observing the firing from this ideal vantage point. At ignition, the fiery thrust of the rocket engines pushed the Atlas skyward with a mighty roar. The brilliant light from the exhaust illuminated the expansive launch area. Initially, the Atlas followed the planned trajectory flawlessly, powering upward into the night sky. Then around 37,000 feet, the point of "max Q" or maximum aerodynamic pressure, the rocket exploded suddenly. "It looked as if an atomic bomb had exploded above us," Glenn observed. This effort at confidence-building by the Mercury program had failed in its objective to reassure the astronauts. NASA continued to upgrade the Atlas and achieved success in transforming it into a reliable rocket, yet, as Glenn later confessed, memories of that ill-fated night launch lingered in his memory when he boarded Friendship 7 on that February morning.[2]

Glenn waited patiently through two hours and 17 minutes of holds before liftoff. When the Atlas engines roared to life, Friendship 7 slowly rose from the pad at Cape Canaveral in a conflagration of fire and smoke, taking a mere 13 seconds to reach transonic speed. At two minutes and 14 seconds into the flight, the booster engines shut down and dropped away. On schedule, the escape tower was jettisoned. As the spacecraft pitched slightly, Glenn had his first view of Earth from orbit. Even as he monitored the controls and made the necessary systems checks, he found time to peer out the cockpit window to capture snapshots of the unfolding scenery from orbit. As Friendship 7 moved down the Atlantic range, the Canary Islands came into view, and he could see the far shores of Africa. The sky above had turned black—an enveloping dome that framed the distant horizon with its ever-shifting hues of blue. As he glanced downward, he saw heavy cloud formations. At this moment, flying in zero gravity, he was moving at 17,544 miles per hour on an orbital path nearly 125 miles above the surface of the planet. The liftoff and early stages of the flight had gone well. The separated Atlas rocket now trailed behind Friendship 7, caught in an erratic tumbling motion and slowly receding from view. Glenn's report to Mission Control that the view was "tremendous" erred on the side of brevity and in no way captured the exhilarating experience of flying in Earth orbit.[3]

The first of what became three orbits turned out to be carefree for the most part. Friendship 7 crossed into Africa above Kano, Nigeria, where NASA had set up one of its global tracking stations to allow for regular communications and precise control of the orbiting capsule. Glenn then flew across the Indian Ocean toward Australia. Now he was in place to view his first sunset. This passage through the night-side phase was followed by a spectacular sunrise. On this fast-moving circumnavigation of the planet the night lasted for 45 minutes. And every five minutes, Glenn shifted his radio communications with a new capsule communicator (capcom) stationed at a tracking station along the orbital path. During this first orbit Glenn reported one strange phenomenon—the appearance of "fireflies," or brilliantly lighted specks, on the outside of the capsule. This peculiar and unanticipated event, subsequently observed during the flight of Scott Carpenter, perplexed Glenn. It was later explained as ice crystals formed from escaping fluids that had frozen on the capsule.[4]

The flight of Friendship 7, however, soon encountered a sequence of real problems, one minor in nature, the other life threatening. Even before the end of the first orbit, Glenn noticed that the automatic attitude control system was malfunctioning, prompting a persistent drift to the right. To remedy the situation, he shifted to manual control and placed the spacecraft on a proper path. A second problem, and one that NASA's capcoms were slow to share with Glenn, was the alarming telemetry suggesting that Friendship 7's heat shield and the compressed Landing Bag Deploy were not attached in the locked position. If this reading were accurate, the critical heat shield was being held in place only by the straps of the retro-rockets. Repeated queries from NASA control concerning the problem were less than candid. Only with time did Glenn realize the scope of the problem, and later he would complain forcefully that he should have been informed. As a veteran test pilot, Glenn was angered at this deviation from protocol. He remarked later that if the NASA controllers thought that any information should be held from a pilot, then that pilot should not be in the cockpit.[5]

The heat shield problem led to the decision to bring Glenn down as soon as possible, to be accomplished after the third orbit. To reduce the risk of losing the heat shield on reentry—a catastrophic event assuring the immolation of the Mercury astronaut—NASA controllers decided to allow the retro-rocket package to remain in place during the fiery passage through the upper atmosphere. The retro-rockets were essential to slow Friendship 7 into a gradual descent into the atmosphere. Mission control speculated that if the fuel in the retro-rockets burned cleanly, there would be no problem—the retro-rockets would be burned away during the descent. This maneuver became the most dangerous part of the mission. Glenn reported that he could see chunks of the straps pass his window as the retro-pack was burned away. The reentry path called for Friendship 7 to follow a long glide across the continental United States and then to splash down in the Atlantic. The capsule landed about 40 miles from its targeted landing zone near Bermuda.[6]

Glenn had been aloft for almost five hours. Orbital flight was a unique experience for these pioneering astronauts, evoking a new visual perspective on Earth. Along the fast-paced orbital journey, he observed the shifting colors of the planet's thin atmosphere—a band of light that reflected brilliant hues of blue and white. He could see clearly the blunt edge of the atmosphere—an extraordinary vantage point denied to other mortals. Glenn, alone in the zero gravity, comprehended fully the atmosphere's enveloping and life-sustaining presence and seeming fragility. "The band,"

Glenn observed, "is extremely bright just as the sun sets, but as time passes the bottom layer becomes a bright orange and then fades into reds, then on into the darker colors, and finally off into the blues and blacks."[7]

Seeing the atmosphere from space also reinforced an appreciation for the atmosphere's role as the shield against cosmic radiation—no life on the blue planet would have been possible without this thin layer of protection. At the time of the Glenn flight—and the earlier Gagarin and Titov orbital missions—a significant amount of speculation still existed about what would happen to a human flying above this atmospheric shield: What dire impact might result from zero gravity, cosmic rays, or even meteor bombardment? Glenn returned without any apparent negative effects, as did his Russian counterparts. The question of human vulnerability to the dangers of outer space would persist throughout the decade of the 1960s.

Glenn's successful mission triggered a vast outpouring of enthusiasm, reminiscent of Charles Lindbergh's reception after his epic transatlantic flight of 1927. The astronaut from New Concord, Ohio, became an American icon overnight. President Kennedy called Glenn shortly after his splashdown and recovery in the Atlantic. After two days of debriefing on Grand Turk Island, Glenn returned home to a tumultuous welcome. Vice President Lyndon Johnson flew down from Washington to escort Glenn back to Cape Canaveral. There he was reunited with his wife, Annie, and his two teenage children, David and Lyn. Welcoming signs at the Cape greeted Glenn, "Welcome to Earth." Later, President Kennedy flew in to greet Glenn. Kennedy and his staff had been glued to the television coverage for the entire saga of Friendship 7, following nervously the tense reentry and recovery phases of the space mission. Received as a genuine American hero, Glenn made a long tour to Washington for a more formal reception with President Kennedy and a ticker-tape parade in New York City. Glenn fit the heroic mold comfortably—he was articulate, charming, and gregarious. Speaking to the NASA community at the Cape, he spoke extemporaneously and caught the mood of the moment: "There is much acclaim for this flight, but it is only one step in a long program. I'd like all of you who worked on it to feel that I am your representative. I'm getting the attention for all the thousands of you who worked on it."[8]

The post-Glenn Mercury flights extended the American presence in Earth orbit. In May 1962, Scott Carpenter in Aurora 7 replicated the Glenn mission, but experienced a splashdown 250 miles distant from the targeted landing site. The following October, Walter Schirra in Sigma 7 sent back the first televised broadcast by NASA from orbit, and he doubled the number of orbits flown by Glenn. The final Mercury mission took place in May 1963. Gordon Cooper in Faith 7 executed the most successful space mission to date, orbiting the Earth 22 times. Mercury represented a great success for NASA.

SOVIET COSMONAUTS AT THE CUTTING EDGE

The dramatic success of John Glenn in Friendship 7 marked an important milestone for Project Mercury and America's manned space program. If the suborbital flights of Shepard and Grissom had been perceived as cautious and telltale signs of NASA's failure to keep pace with the Soviet space program, Glenn's three orbits affirmed that the United States was now competitive with the mysterious Chief Designer and his minions at Baikonur. But the triumphal Glenn saga still remained modest when compared with German Titov's impressive 17 orbits in August 1961. The final Mercury missions by M. Scott Carpenter and Gordon Cooper narrowed the gap. Yet the Soviet Union still occupied the pole position in the high stakes "space race"—and at Baikonur, plans were in place to reassert Soviet leadership in the competitive world of space exploration.

The Vostok launches gave clear evidence that the Soviets were on a path to establish some new benchmarks in manned spaceflight. Following the 1961 Gagarin flight were a total of six Vostok missions through June 1963. Broadly considered, the Soviet effort mirrored the same goal of Project Mercury—the urgent priority to perfect the baseline technology needed to sustain a long-term program of space exploration, culminating in a lunar mission. The first benchmark Vostok spectacular occurred on August 11-12, 1962: Vostok 3 with cosmonaut Andrian Nikolayev and Vostok 4 with cosmonaut Pavel Popovich were launched into orbit on successive days. This extraordinary scene of two spacecraft in orbit simultaneously struck many as an important benchmark—and concrete evidence of the Soviet Union's muscular space program. The two new Vostok cosmonauts were polar opposites in personality. Nikolayev, a former lumberjack before he became an air force pilot, was quiet, disciplined, cool in any emergency, and highly respected by his superiors for his technical skills. Popovich, a Ukrainian by birth and an experienced test pilot, was the same age as Nikolayev (32 years old), but he had a more engaging personality, being an extrovert by nature. Their spacecraft did not possess the capability to maneuver or dock, but scripted launches from Baikonur placed them in close parallel orbits, passing within three miles of each other. This rendezvous in space represented a remarkable feat of coordination. The Soviets not only executed the first group flight, but also established a new record for duration in space: Nikolayev completed 64 Earth orbits, Popovich another 48.[9]

Popovich reported at one juncture that he made visual contact with Nikolayev's Vostok 3, describing the sister spacecraft as a "small moon." While in orbit, the two spacecraft followed identical routines with both cosmonauts eating meals and sleeping in identical time slots. They were able to maintain regular radio communication with each other, notwithstanding the occasional static and noise. Television cameras on board each spacecraft broadcast home a chronicle of life in space. Onlookers acquired a vivid sense of weightlessness through images of floating pencils and logbooks. Each cosmonaut ate four times a day; the food in plastic tubes consisted of meat cutlets, roast veal, chicken, pastries, and sweets. There was a systematic program of photography, and the cosmonauts exercised regularly. While both missions were highly automated, each man experimented with the manual control system to alter the attitude of his spacecraft.[10]

The nausea experienced by Titov in Vostok 2 led to the use of a special coded language for the cosmonauts to alert Baikonur of any symptoms of "space sickness." Any severe crisis would trigger an early reentry. For such an emergency, Popovich had been instructed to use the code phrase "observing thunderstorms." As it turned out, Popovich was spared any nausea or disorientation during the entire mission, but he carelessly used the code phrase while describing real thunderstorms he observed over the Gulf of Mexico. At Baikonur, his words "observing thunderstorms" was interpreted to mean that he was suffering severe motion sickness. When asked how he was feeling, Popovich quickly realized his ill-chosen words. When he attempted to reassure Baikonur that he was feeling fine, his reassurances were greeted with skepticism. Mission Control ordered the cosmonaut home on his 48th orbit.[11]

American reactions to the Vostok group flight were often intense and in some quarters accompanied by no small amount of apprehension. "I can remember," NASA official Robert Seamans wrote in his memoir, "that *Aviation Week* carried the story soon after it occurred, and there was much speculation as to Soviet intentions." The seamless ties in the Soviet Union between the space program and the military prompted considerable fear, Seamans wrote: "Were they conducting this dual maneuver to gain experience for their own exploration, or were these tests a prelude for Soviet inspection and possible interception of U.S. satellites?" The Cold War had cast a long shadow over the space missions of both countries.[12]

A second group flight by Vostok 5 and Vostok 6 followed in June 1963, but this time with a surprising new twist—one of the orbiting cosmonauts was a woman. This bold move scored another first for the Soviet Union. Vostok 5 with Valery F. Bykovskiy inaugurated this space spectacular, reaching Earth orbit on June 14. The launch was trouble-free, but the resulting orbit for Vostok 5 turned out to be lower than planned. As a consequence, the orbit of the spacecraft was destined to decay rapidly, an eventuality that would lead to increased cabin temperatures. These conditions predetermined that his journey into space would be cut short, to conclude in five days. Bykovskiy found his view from space exhilarating; he saw landmarks including Leningrad, the Nile River and Cairo, the fjords of Norway, and mountain ranges. At night, he observed violent lightning storms above South America and the night-lights of major cities below his orbital path. Television coverage of Bykovskiy was broadcast in the West.[13]

Two days later, on June 16, Vostok 6 reached orbit with Valentina Tereshkova on board. Her call sign was Chaika ("seagull"). Tereshkova's arrival in Earth orbit created a global sensation and quickly overshadowed Bykovskiy's flight. In the mind of General Nikolai Kamanin, then in charge of cosmonaut training, Tereshkova represented the best female candidate for the historic Vostok 6 mission; in his words, she was *povedeniye* ("a moral exemplar") and possessed *vospitannost'* ("good breeding").[14] Tereshkova was 24 years old, the oldest in a small cadre of female cosmonaut trainees; the group included Irina Solovyeva, Valentina Ponomareva, and Zhanna Yerkina. The choice of Tereshkova reflected the established pattern of favoring those who clearly embodied a communist pedigree and a compliant personality. She received the nod in favor over the more independent-minded and outspoken Ponomareva, a trained scientist, who by some accounts was the most outstanding female candidate.[15]

Given the secrecy surrounding all aspects of the Soviet space program, decades passed before it was learned that Tereshkova's sojourn in space had been a difficult one. Fellow cosmonaut Aleksei Leonov observed that while Tereshkova never lacked courage or failed to perform her tasks as assigned, she did not respond well to space travel, suffering a severe bout of disorientation and at times failed to respond to communication prompts from the ground. Her condition during the flight would remain a matter of speculation in the years that followed. Khrushchev, who had called for the launch of the first woman cosmonaut, took the occasion to praise Tereshkova's achievement as a sign that "men and women are treated equally in the Soviet Union." The fact that Tereshkova had piloted Vostok 6 marked another breakthrough, but Korolev and others remained cool to the idea of assigning women future flying roles, at least in the near term.[16]

In 1964 the Soviets faced the serious challenge of NASA's Project Gemini. The spacecraft built for Gemini were designed for a two-man crew, launched into orbit atop a Titan II booster. The Gemini configuration was larger and more sophisticated than its Mercury predecessor. The ultimate Soviet answer to Gemini was to be the Soyuz, which was still under development. The Soyuz, as with its Gemini counterpart, was a spacecraft designed for multiple crews, orbital maneuvering, rendezvous and docking, and flights of up to 30 days' duration. The advanced Soyuz consisted of a command module (serving as the ascent and descent capsule); an orbital work module (a spherical component offering spacious room for work and an air lock for space walks); and an instrument and systems module (containing propulsion, electronics, and communications systems). To the chagrin of Korolev, however, the Soyuz would not be available in the near term; it would not be flown until 1967.[17] Consequently, Korolev began to think of an interim solution, one that would adapt the veteran Vostok spacecraft for a new role as a multiple-crew space vehicle.

With urgent, cruel deadlines, Vostok was transformed into the Voskhod ("sunrise") spacecraft and quickly deployed for a new series of risky space ventures. Two critical tests of the Voskhod took place in October 1964 and March 1965. The new spacecraft was a radical modification of the older one-seat Vostok with a gutted and redesigned interior to accommodate additional crew. As such, Voskhod did not flow out of any logical and incremental process of design; instead it was a desperate attempt to maintain competitiveness with the United States. Khrushchev played a key role in the unfolding drama, thirsting for more space spectaculars and seeking to find an effective means to preempt the Gemini program.

Cosmonaut Leonov claimed that he approached the new Voskhod with great interest. His inspection of the spacecraft was conducted in a bizarre context of secrecy. Leonov was among a group of cosmonaut trainees Korolev had summoned to his design bureau, OKB-1. The Chief Designer gave specific instructions for the cosmonauts to appear in civilian clothes. To prepare for the top-secret conclave, Leonov and his fellow trainees hurriedly purchased new suits. But, to their surprise, when they boarded the bus for the design bureau, they were dressed in roughly identical outfits, prompting an outbreak of laughter: "We had simply exchanged one uniform for another," Leonov recalled. "We were all wearing the same brick-colored overcoat, tailored by expensive Italian designers." Once the cosmonauts entered the design bureau, they discarded their expensive civilian garb for white smocks and covers for their shoes.[18]

Leonov was not surprised by this elaborate exercise in secrecy. On one level, Korolev's approach mirrored a political culture where secrecy was normal. The penchant for strict secrecy also reflected Korolev's decrees for a highly disciplined lifestyle, a set of values he imposed on his family and workers.

Korolev's movements were not routinely forecast or widely known. When on vacation in a sanatorium along the Black Sea, he enjoyed unrivaled privacy, living in quarters sealed off for his personal and exclusive use. Though he was denied proper public credit for his achievements, Korolev's clandestine life did offer an exemption from certain petty tasks associated with public affairs. His workload was daunting and all consuming. Leonov well remembered his toughness: "Korolev had the reputation of being a man of the highest integrity, but also extremely demanding. Everyone around him was on tenterhooks, afraid of making a wrong move and invoking his wrath. He was treated as a god. . . . He did not suffer fools gladly. He had the ability to silence a person with the smallest gesture of a hand."[19]

Seeing the new Voskhod from a distance, Leonov and his fellow cosmonauts thought they were looking at a Vostok space vehicle, but this initial impression proved false. Closer examination revealed substantial differences between the two spacecraft. The most striking change was the attachment of an air lock to the Voskhod, which Korolev explained in nautical terms: "Every sailor on the ocean liner must know how to swim. Likewise, a cosmonaut must know how to swim in open space."[20] Looking at the radically redesigned interior, Leonov recognized that Voskhod spacecraft would accommodate two or three cosmonauts in future missions. Other modifications had occurred in instrumentation and in the placement of control panels. It was a new spacecraft, configured for soft landings, but retaining much of the DNA of the old Vostok model.[21]

The first launch of the new spacecraft came on October 12, 1964. Voskhod 1 lifted from its launch pad at Baikonur with a crew of three—command pilot Vladimir Komarov, an air force officer and member of the original corps of cosmonauts, accompanied by two non-flying specialists: Konstantin Feoktiskov, a scientist, and Boris Yegorov, a physician. To allow freedom of movement, the crew wore no space suits. The three seats in metal cradles were set perpendicular to the old Vostok ejection seat position, an awkward configuration since the crew could read the instrument panel only by craning their necks. Only one safety add-on adorned the new Voskhod design, a solid retro-rocket package placed on the nose of the spacecraft. The mission remained a high-risk endeavor. Any emergency at liftoff or an explosion on the pad would mean death for the crew.[22]

Television cameras on board offered periodic coverage of the Voskhod 1 crew during their epic journey. They made a total of 16 orbits over 24 hours, 17 minutes in space. The Soviet feat of catapulting a three-man crew into Earth orbit made a singular impact on observers in the United States, where many attributed exaggerated prowess to the Soviet space program. The engineering details behind Voskhod 1 remained largely hidden. The Soviet space triumph, as subsequent events would reveal, was more apparent than real. For the moment, though, the impression that the Soviet Union enjoyed a wide lead over the United States in space technology remained fixed, even reinforced.

One of the great ironies of the Voskhod 1 mission was Nikita Khrushchev's removal from power even as the cosmonauts were aloft. The Soviet leader was toppled by a cabal led by Leonid Brezhnev and Aleksei Kosygin. Discontent with Khrushchev's rule had festered for many years; the plotters viewed his erratic and adventuresome ways as dangerous. In the aftermath, Brezhnev assumed the post of First Secretary of the Communist Party. The removal of Khrushchev cast doubts on the future of the Soviet space program. But Brezhnev did not reverse the high-profile campaign to engage in space spectaculars to enhance the prestige of the Soviet Union. Nor did he abandon, in any formal way, the idea of competing with the United States to be the first to land on the moon.

A WALK ON THE WILD SIDE

On March 18, 1965, Voskhod 2 began one of the most remarkable space adventures of all time. This second Voskhod mission would be remembered for the first "walk in space." That epic achievement, however, was just one episode in an extraordinary story of human courage and survival. The two-man crew of Voskhod 2 consisted of Pavel Belyayev, the command pilot, and Alexei Leonov, the cosmonaut chosen by Korolev for the first "swim" in outer space. Both men were experienced air force pilots and highly regarded for their boldness, flying experience, and technical skill. Their exploits that March qualified them for inclusion in any Russian version of *The Right Stuff.*

Just before his fall from power, Khrushchev had approved the idea of a space walk, seeing it as a stunning way to overshadow the Americans' newly inaugurated Gemini program. For Korolev, the high-risk Voskhod 2 mission involved more than just a propaganda stunt. Any future program of space exploration, he felt, would require the perfection of techniques for extravehicular activity (EVA). To achieve this end, Soviet engineers fitted Voskhod 2 with a special airlock to stage a space walk. This cylindrical-shaped appendage could be extended and then retracted with the use of inflatable booms. The device offered a cosmonaut a secure corridor to reach safely the void of outer space.

Belyayev and Leonov were close friends who had undergone a rigorous training regimen in preparation for their mission. This training program, carefully monitored by Korolev, called for the simulation of every phase of a space walk. The actual walk would last close to 15 minutes, but the techniques for moving a cosmonaut in and out of the capsule proved to be daunting. The airlock, though experimental and untested, appeared to be a workable, even clever, design. For both cosmonauts to acquire the sensation of weightlessness, they had been flown on a modified Tu-104 jet transport in a series of parabolic arcs. There were still many unknowns, since a space walk had never been tried, prompting controllers at Baikonur to speculate about a number of emergencies: If, for example, Leonov lost consciousness in space, how would Belyayev rescue him?[23]

Just three weeks before the launch of Voskhod 2, an unmanned prototype was shot aloft from Baikonur, only to explode when the spacecraft's onboard self-destruct system was triggered accidentally. This reversal placed considerable strain on Korolev and his team; they were now down to one Voskhod spacecraft. A replacement vehicle for testing would not be available for a year. Korolev was inclined not to wait. Asking his cosmonauts for their opinion, he reminded them of the inherent risks in the launch of the one remaining Voskhod: "It is up to you," he warned. "I cannot tell you what you should do. There is no textbook answer. Nothing can prepare you exactly for what you will experience on your mission." Korolev then appealed to their competitive instincts, reminding them that the American astronaut Edward H. White was scheduled for a Gemini space walk in May. According to Leonov, this factor swung the scales: "That night in late February 1965," he recalled, "we were full of self-confidence. We felt we were invincible. Despite the risks, we said, we were ready to fly."[24]

On the morning of March 18, Belyayev and Leonov rode the Voskhod 2 into orbit without difficulty. The mission was planned for one day, so the crew moved quickly to prepare for a space walk on the second orbit. At this juncture, Voskhod 2 was traveling at nearly 18,000 miles per hour in an orbit of 103 miles at its perigee and 295 miles at its apogee. Mission control followed the preparations for the EVA with intense interest, finally giving the go-ahead. Belyayev then began pumping air into the rubber tubes of the inflatable air lock; nearby in the cramped interior of the capsule Leonov strapped himself into his bulky spacesuit. He would have 90 minutes of oxygen for the space walk. Entering the fully extended air lock, he then waited until the pressure in the chamber equaled the zero pressure of outer space. The hatch was then opened.

Leonov moved slowly to extricate himself from the narrow chamber of the air lock. "What I saw as the hatch opened took my breath," he recalled. "Night was turning to day. The small portion of the Earth's surface I could see as I leaned back was deep blue. The sky beyond the curving horizon was dark, illuminated with bright stars as I looked due south toward the South Pole." Leonov then edged cautiously outside the air lock, holding onto a special rail, a move that took just a few seconds. Looking around from this unique vista, he realized he was then flying over the Mediterranean. On the horizon he could see the entire Black Sea and to one side, Greece and Turkey. In the far distance, he could see faintly the Caucasus Mountains and the Volga River. To the north, even the Baltic Sea was in view. Leonov remembered clearly the next moment: "With a small kick, as if pushing away from the side of a swimming pool, I stepped away from the rim of the airlock. I was walking in space. The first man ever to do so."[25] Televised images of this dramatic moment were viewed by millions in the Soviet Union.

This sojourn in the void of outer space was, by design, short-lived. Almost immediately Leonov realized that his walk would not be without complications and dangers. He first noticed a sharp increase in the temperature of his space suit. This discomfort was accompanied by an inability to manipulate his hands effectively to take photographs with his handheld camera. As he began to move back toward the airlock for his reentry into the capsule, he noticed that his spacesuit had become deformed. To add to the problem, he realized that his feet had pulled away from his boots and his fingers from the attached gloves. Leonov now realized his peril—he could not return feet first into the airlock as scripted.[26]

Acting alone and in extreme danger, Leonov moved instinctively to save himself. He decided to turn over and enter the airlock headfirst. In an improvised emergency maneuver he moved carefully, twisting and turning until he gained access. Only with great effort did he wriggle into a position where the outer hatch could be closed again. To achieve reentry into the capsule, he faced yet another challenge: what to do with his inflated and misshapen space suit? Again, Leonov performed a dangerous maneuver. He chose slowly to bleed off some of the oxygen from his pressurized suit, made possible by a valve in the suit's lining. This was the only way to reduce the pressure in his space suit. It took great physical effort to perform these movements. Leonov later explained his heroic struggle: "At first I thought of reporting what I planned to do to Mission Control, but I decided against it. I did not want to create nervousness on the ground. And anyway, I was the only one who could bring the situation under control." Facing extreme peril, Leonov remained cool, devised his own path to safety, and narrowly escaped death.[27]

His alarmed crewmate Belyayev warmly welcomed an exhausted Leonov back into the safety of the capsule. Leonov was drenched in sweat. Later he learned that during this physical ordeal he lost 12 pounds.

None of this drama was televised to the millions of onlookers at home. Once the Voskhod 2 spacecraft appeared to be in some difficulty, higher authorities in Moscow decided to end all radio and television coverage. This was done Soviet-style, abruptly and without any explanation. As a substitute, the airwaves were filled with solemn cadences of Mozart's Requiem. The choice of music evoked fears for the crew of Voskhod 2, since the Requiem was typically played at state funerals. Days would pass before the fate of Belyayev and Leonov would be known to the public. The Soviets had executed yet another space spectacular, but it had nearly been a disaster.[28]

Both men soon discovered that there were other challenges yet to come. The oxygen pressure in the capsule started to increase at an alarming rate, owing to a breakdown in the environmental control system. The increased pressure meant that an errant spark from the electrical circuitry could ignite an explosion. This particular danger, however, lessened with time. The beleaguered crew then learned that the automatic guidance system had become inoperative, which meant that the primary retro-rocket system had failed on Voskhod 2. This required the use of manual controls for reentry. The passage through the upper atmosphere was complicated further by the fact that the service or landing module failed to make a clean separation, sending the spacecraft into a series of gyrations. This dangerous spinning ceased at around 60 miles when the cosmonauts broke free of their orbiting module. Now off course, the Voskhod 2 made its final descent nearly 1,200 miles away from the planned recovery site, landing in a remote area near Perm in the Ural Mountains.[29]

Leonov later recalled the reassuring "sharp jolt" that signaled the deployment of the drogue and the landing chutes. The capsule began its slow drop to the ground, being tossed gently in the wind. The landing engine ignited on schedule, breaking the momentum. Belyayev and Leonov, strapped tightly in their metal cradle seats, then passed into a dense forest area of the taiga, landing in a deep snowdrift. Opening the hatch proved complicated, since it was now blocked by a birch tree. After an awkward interlude, the cosmonauts extricated themselves from the jammed hatch. For the first time they breathed fresh air and felt the extreme cold of the taiga. With the use of an emergency transmitter, Leonov began sending distress signals in Morse code to mission control as darkness set in.[30]

Finding themselves stranded in the forest was reminiscent of a plot in a medieval Russian fairy tale—a dark and foreboding forest filled with wild bears and wolves. It was spring, the mating season when both of these animals were extremely aggressive and not welcoming to any chance visitors from Baikonur. The sight and the howling of a roaming pack of wolves prompted a decision by the cosmonauts to spend the night inside their capsule. Leonov had brought along a pistol, and he had an ample supply of ammunition, but fate spared them any threatening encounter. Their sojourn in the forest would last two days before recovery. Once Mission Control determined the location of their lost cosmonauts, they dispatched rescue helicopters to retrieve them—a prolonged effort given the remote locale and the need to carve out a landing strip in the forested area. On March 21, Belyayev and Leonov were evacuated to Perm, ending one of the most arduous chapters in space history.[31]

PROJECT GEMINI ADVANCES THE GAME

NASA announced Project Gemini in December 1961. The goal of Gemini was to develop and refine the necessary techniques for a lunar mission. During the 20-month history of Gemini, ten manned missions would be launched into Earth orbit. As the logical successor to Project Mercury, Gemini showcased a new generation of spacecraft. Gemini carried two astronauts in a capsule weighing twice as much as its single-set Mercury counterpart. It boasted new ejection seats to provide for emergency escape for the astronauts, replacing Mercury's escape tower. As with the Voskhod series, the Gemini two-man missions gathered physiological data on the impact of prolonged spaceflight on astronauts. Another primary goal was to execute a walk in the hostile environment of outer space. In addition, Gemini would develop and test rendezvous and docking in space, a task viewed as essential to any moon mission. In the beginning the Gemini spacecraft design was simply a scaled-up version of the Mercury—one large enough to carry two astronauts. This conception was articulated by NASA's Space Task Force in association with the McDonnell Aircraft Corporation, the builder of the Mercury spacecraft. This basic design, however, developed into a more sophisticated vehicle, using advanced fuel cells rather than batteries, allowing for extended space journeys of two weeks or more.[32]

With its configuration for two astronauts and the added equipment, the Gemini capsule was too heavy to be carried into orbit by the Atlas missile employed in Project Mercury. NASA selected a man-rated version of the Air Force's second ICBM, the powerful Titan II, for the Gemini program. One of the four Titan test launch pads at Cape Kennedy was modified for the Gemini flights. The Titan II used hypergolic fuel, composed of chemicals so reactive that they burst into fire upon contact with each other, without needing an ignition source. A second vital ingredient of Gemini was using the Atlas missile to orbit an Agena second stage as a target for rendezvous and docking experiments. This ambitious program soon encountered difficulties and delays, requiring more time and effort than NASA had anticipated. Nor did Gemini's problems end there: The Titan II suffered from a condition known as longitudinal oscillations, referred to as the "pogo effect." In addition, the Gemini's fuel cells developed leaks, requiring a redesign and adaptation. Program costs rose with the many delays, eventually tripling to more than one billion dollars.[33]

The first unmanned Gemini launch atop a Titan II fell behind schedule, being delayed 11 months until April 1964. However, this initial mission, Gemini 1, completed 64 orbits. This and a second unmanned test validated the Titan II's readiness to carry astronauts as well as the missile-capsule interface. In March 1965, Gemini 3 became the first manned mission. Crewed by Gus Grissom and John Young and dubbed the *Molly Brown,* it completed three orbits in the five-hour flight. Using the Gemini's Orbit Attitude and Maneuvering System (OAMS), a propulsion system composed of 16 small engines, the astronauts were able to alter their orbital flight path, an essential capability for a future space rendezvous.

In June, the Gemini 4 mission achieved a number of important milestones. The spacecraft stayed in orbit for four days, far longer than the Mercury flights and equivalent in time to a projected journey to the moon.

The stellar achievement of the mission, however, was the 21-minute space walk by Ed White, the first EVA for the United States. White had been one of nine new astronauts selected by NASA in September 1962 in the first addition to America's astronaut corps following the original Mercury astronauts in 1959. Born in San Antonio in 1930, White flew Air Force fighters in Germany and later served as a test pilot.[34]

With his colleague James A. McDivitt flying Gemini 4, White began his adventure in space with the necessary depressurization of the Gemini cabin. The hatch door was then opened as the spacecraft sped above the Indian Ocean. At the start White faced some difficulty in opening the hatch, but this awkward moment passed quickly. He then moved in a deliberate manner, making use of his "Zip Gun," a double-barreled, compressed-air maneuvering gun, to propel himself into the void of space. He was exhilarated as he moved about in the ocean of space, secured only by a tether. As he maneuvered and took in the dramatic vista of Earth from orbit, he was flying upside down at 17,500 miles an hour. White shared his excitement: "This is the greatest experience; it's just tremendous. . . . Right now, I'm standing on my head, and I'm looking right down, and it looks like we're coming up on the coast of California. . . . There is absolutely no disorientation associated with it." During the walk White experimented with his unique handheld gun to travel a distance of about 15 feet and moving out and above the capsule, discovering in the process that he had flown higher than planned. He then corrected his position vis-à-vis the capsule with some short bursts, reporting to Mission Control that the gun operated well. He went through a few additional body turns using the gun, and then—to his great disappointment—the Zip Gun became empty. The spacewalk had been brief, dramatic, and executed without any mishap. When it came time for the inevitable return to the capsule, a reluctant White told McDivitt: "It's the saddest moment of my life."[35]

Earlier in the flight, one maneuver by the Gemini crew had not gone as smoothly. They attempted to rendezvous with the second stage of the Titan II, then in orbit a few hundred feet behind them. The Gemini crew successfully turned their spacecraft around to face the spent booster and then advanced toward it, using the capsule's thrusters to control their movement. Yet they failed to close the gap between the booster stage and Gemini 4. McDivitt recognized that the mission's first priority, the EVA, could be endangered by continued expenditure of the capsule's limited fuel for orbital maneuvering and ended further attempts at rendezvous. This was the first time that a link-up of one spacecraft with another had been attempted in space.[36]

Later Gemini flights did achieve considerable success in perfecting rendezvous techniques while at the same time demonstrating flexibility and innovation in responding to unexpected situations. A prime example of this occurred when an Agena upper stage with which Gemini 6 was to rendezvous failed to reach orbit. Lacking a target, the Gemini 6 launch was cancelled and replaced with an ambitious new assignment for its crew: an orbital rendezvous with Gemini 7. Because the rendezvous would require two Gemini launches within days of each other, and only one Gemini launch pad was available, the challenge facing NASA was to launch Gemini 7 and then clean, refurbish, and prepare Pad 19 for the launch of Gemini 6 (renamed 6-A for the new mission) in an unprecedented eight days. Normally, two months were allotted for this task, which included repairing the considerable damage to the launch pad from the superheated exhaust of the Titan II's engines as the rocket lifted from the pad. Other essential functions were the erection and testing of the Titan II, mating it with the capsule, and checking out the complete assembly. Gemini 7, carrying Frank Borman and James Lovell, Jr., orbited on December 4, 1965, for a 14-day mission. The refurbishment and launch preparations at Pad 19 were completed within the allotted time frame, though a last-minute problem delayed the Gemini 6-A launch until December 15.[37]

The orbital maneuvering for the rendezvous took several hours. When Gemini 6-A, carrying Wally Schirra and Tom Stafford, was slightly below and 37 miles behind Gemini 7, Stafford flew his capsule as close as two feet to Gemini 7. Each capsule was able to maneuver in turn to inspect the other from all directions. When the rendezvous was completed, the capsules "parked" about 30 miles apart while the crews slept. For the first time two vehicles had maneuvered to meet in space.

That mission was not the end of Gemini's triumphs, or, for that matter, of the program's problems. In March 1966, Neil Armstrong and David Scott set off in Gemini 8 to achieve the next step following space rendezvous: orbital docking. Just minutes after closing in on the Agena target and successfully docking with it, however, the two merged spacecraft began spinning uncontrollably. Scott separated the Gemini capsule from the Agena, but the spinning increased to a full revolution every second. Concerned that they might become too dizzy to regain control, Scott radioed NASA Mission Control: "We have serious problems here. We're tumbling end over end. . . . We've disengaged from the Agena." The crew was able to use their reentry control system to stabilize the spacecraft, but they had to make an emergency return to Earth, using a "secondary" landing area near Okinawa. The crew was uninjured, but it had been a close call.

Armstrong, like John Glenn an Ohio native, would go on to achieve worldwide fame as the first human to step onto the moon. He joined the astronaut corps in 1962, one of nine members of NASA's second astronaut class. Armstrong had learned to fly at 15 years of age and had a student pilot's license before he earned his driver's license. He later flew combat missions for the Navy during the Korean War and piloted the legendary NASA X-15 rocket-powered research plane, achieving speeds of 4,000 miles per hour.[38]

Like Armstrong, nearly all of the Gemini astronauts were selected from NASA's Group 2 and Group 3 astronaut classes, who joined the original Mercury Seven in 1962 and 1963, respectively. Nine were added in 1962 and 14 in the following year. Of the original Mercury astronauts, only Grissom, Schirra, and Cooper flew Gemini missions. But many of the members of Group 2 and Group 3 would go all the way—to the moon.

In July 1966, Gemini 10's John Young and Michael Collins succeeded in docking with their Agena target and holding both spacecraft steady. They fired the Agena's propulsion system, taking the combined spacecraft's orbit out to 547 miles at its apogee, a new record for manned flight. "That was really something," Young reported after the Agena burn. "When that baby lights, there's no doubt about it." Two months later, Gemini 11 reached

850 miles, powered by the Agena stage it had docked with. Astronaut Charles "Pete" Conrad described what he could see from that height: "It's fantastic. You wouldn't believe it. I've got India in the left window, Borneo under our nose, and you're [the controllers in Australia with whom he was talking] in the right window. The world is round." At their height, the horizon-to-horizon view encompassed nearly 5,000 miles.[39] With these successful docking missions, American astronauts had demonstrated their mastery of the complex, vital science of orbital mechanics, the process by which one orbiting object catches up with another in space to achieve rendezvous and docking.

The NASA Gemini simulator played a key role in the training regimen of the astronauts by providing a vehicle for terrestrial practice in precision maneuvering and complex orbital calculations for rendezvous. The other key ingredient for success was the onboard computer. Gemini was the first American spacecraft to be fitted with a computer, which performed essential calculations for the execution of precise maneuvers in space. This necessity arose because all objects orbiting the Earth must resist the force of gravity to do so. That force varies with the distance from Earth, so that satellites closer to the planet need to orbit at a greater speed in order to resist the Earth's gravitational pull. Orbiting objects at a greater distance away resist the planet's gravitational pull more and therefore travel slower. Because of this link between height and speed, an astronaut needing to slow down to carry out an orbital rendezvous would have to use his rocket engine to kick himself into a higher orbit. Speeding up required using the engine as a brake to drop into a lower orbit. A hypothetical "error" by an astronaut, such as increasing his speed while trying to overtake a target vehicle, would result in a higher orbit at a reduced velocity. The target would pull away until it disappeared beyond the Earth's horizon; it would then circle the planet, eventually reappearing behind and below the astronaut's capsule.[40]

In November 1965, the highly successful Project Gemini series came to an end with the launch of the twelfth mission. Crewed by Jim Lovell, Jr., and Edwin "Buzz" Aldrin, Gemini 12 became one of the most spectacular outings in the program. During a four-day flight, the final Gemini mission featured a total of 5.5 hours of EVA by Aldrin, encompassing three separate walks. Aldrin utilized handrails, foot restraints, and waist tethers during his EVA, and reported that underwater training simulating EVA had been very helpful in preparing him for the spacewalks. A member of NASA's third (1963) astronaut class, Aldrin attended West Point and then flew F-86 Sabres for the Air Force in Korea, shooting down two Russian-built MiG-15 fighters. Before becoming an astronaut, he had earned a Ph.D. from the Massachusetts Institute of Technology with a dissertation focusing on manned space rendezvous. He made effective use of his academic studies at NASA, working tirelessly to develop solutions to the highly complex problems of orbital rendezvous and docking.[41]

Project Gemini made major contributions to the emerging American space program in the mid-1960s, accumulating 24 million miles of manned spaceflight experience, the equivalent of more than 50 round-trip journeys to the moon. Gemini represented a dramatic advance over the cautious up-and-down spaceflights of Project Mercury. Americans were now making prolonged stays in outer space, up to 14 days. All these impressive achievements set the stage for Project Apollo and the upcoming voyages to the moon.

A PARALLEL SPACE RACE—THE ROBOTIC MISSIONS

Even as America prepared its astronauts for a lunar mission, a parallel space race unfolded. Both the Soviet Union and the United States shared the realization that much remained to be learned before humans could set foot on another celestial body. For example, no one knew what the astronauts or cosmonauts would encounter when they landed on the moon. How much lunar dust would be kicked up, and would it interfere with the astronauts' visibility? Was the surface stable enough to support astronauts and their spacecraft, or would they sink into lunar dust? What about radiation in space?[42]

The only way to answer these and many other questions was to send unmanned robotic scout vehicles ahead of humans. The key precedent for such an endeavor was the early Soviet Luna moon probes, which, along with the American Pioneer series, represented the first attempts to travel beyond Earth orbit, starting in 1958. Their success record, while decidedly mixed, did yield some successes for both nations, including flybys of the moon, lunar orbits, probes that hit the moon, and others that went into orbit around the sun. These flights helped both nations develop a logical order of progression for space exploration in the 1960s. The tasks were categorized by complexity and sophistication into flyby, orbiting, and landing. As the names imply, flyby probes whizzed by their target planet, or the moon, accumulating basic data that could be used to allow later spacecraft to orbit around the moon or a future planet. These vehicles captured more intensive information during repeated passes over their celestial targets. That data, in turn, helped planners to evaluate the best places to land. In the case of the moon, at least, humans were sure to follow.[43]

This was the exploration model that NASA adapted for its unmanned moon explorations, an essential precursor to sending astronauts to the lunar surface. First in line was Ranger, designed and constructed by NASA and California Institute of Technology's storied Jet Propulsion Lab (JPL). Ranger's mission was to send detailed television pictures and other data to Earth as it approached the moon and then crashed into its surface at high speed. The early Ranger missions did not go well: Rangers 1 and 2, test models launched in 1961, were stranded in low-Earth orbit when Agena upper stages, which would boost them toward the moon, failed to restart. Ranger 3 fared no better in early 1962 on a mission representing the first U.S. attempt to achieve impact on the moon; the engines of its Atlas-Agena booster fired for too long. As a result the Ranger missed the moon by 20,000 miles, ending up in orbit around the sun.[44]

More failures followed, although Ranger 4 did at least crash into the lunar surface in April; unfortunately, it lost power en route and sent back neither images nor data. It nonetheless marked the first time the U.S. had hit the lunar surface, three years after Russia's Luna 2 had done so. After Ranger 5 also missed the moon that fall, NASA voiced serious concerns about JPL's competence to carry out the Ranger mission. An investigation into the failures resulted in major changes that reduced each mission's goals. For a time, NASA was concerned that the Apollo landings might be delayed by Ranger's failures to return photos of potential lunar landing sites. Success finally came in July 1964, when Ranger 7 sent back more than 4,000 clear and impressive images on videotape and film before crashing into the northern rim of the Sea of Clouds. The final picture, taken from 1,500 feet above the lunar surface, revealed craters as small as three feet across. Three additional Rangers each returned thousands more photographs on similarly flawless missions through March 1965. JPL's reputation had been redeemed, and the data sent back by the Rangers would be used to select landing sites for Surveyor, the next step in reconnoitering the moon.[45]

Surveyor was a 600-pound three-legged triangular aluminum structure loaded with instruments. It carried a solid-propellant engine at its base to enable it to make a "soft" landing on the moon. Once there, it would use a sophisticated imaging system to take and transmit close-up color photos of the lunar landscape back to Earth. JPL managed this program as well, though its development took a backseat to the myriad problems encountered by Ranger. Unlike Ranger, though, Surveyor scored a spectacular success on its first launch, a feat made even more impressive by the major delays and difficulties with its Atlas-Centaur launch vehicle. (The Centaur upper stage was powered by new and powerful engines fueled by liquid oxygen and liquid hydrogen to give it heavy-lift capability, but it lagged years behind schedule. Concurrent development of a new launch vehicle and a new lunar probe, both unproven, greatly increased the odds of failure.) Surveyor 1 landed on the Sea of Storms on June 2, 1966, and sent back more than 11,000 high-quality color photos, along with seismological data and other information, over a six-week period. The photographs, the first taken on another celestial body by a U.S. spacecraft, showed a desert-like landscape strewn with rocks.[46]

The American robotic probe was not, however, the first spacecraft to achieve a soft landing on the moon or to demonstrate that a spacecraft could land on the lunar surface without sinking into it. Those honors had gone to the Soviet Union four months earlier, when Luna 9 made a soft landing in the same Sea of Storms area as Surveyor 1. Luna 9 opened the metal petals forming the spacecraft to stabilize itself on the lunar surface and soon began to transmit a series of television pictures, which were assembled into a panoramic view of the lunar surface, including the horizon nearly a mile away. However, Surveyor 1's images were of a much higher quality; in addition, at least five prior Soviet attempts to achieve a soft landing on the moon between April 1963 and the end of 1965 had failed. Each of the five had either impacted at high speed or missed the moon completely.[47] Surveyor 2 crashed into the moon, but Surveyor 3 performed perfectly, landing on April 19, 1967, at the edge of a shallow crater, also in the Sea of Storms. It sent back more than 6,300 photographs and used its robotic arm, or claw, to dig trenches in the lunar surface and examine its consistency. Surveyor 3's achievements would yield the landing site used by Apollo 12 in November 1969. Three more Surveyors performed superbly before the program's final mission in January 1968, demonstrating the feasibility of manned landings and explorations of the lunar surface.[48]

The final space probe in NASA's pre-Apollo exploratory triad was the Lunar Orbiter, managed by the Langley Research Center. Its mission was to photograph the moon so that accurate maps could be prepared and the best landing sites for the coming Project Apollo landings selected. Five Lunar Orbiters, each weighing 800 pounds, were launched between August 1966 and August 1967. All were highly successful, returning 200 pictures each. In addition to mapping nearly all of the lunar surface, Lunar Orbiters also mapped the moon's gravitational field and provided data on radiation levels on and around the moon. Once in lunar orbit, the robotic probe used its velocity-control rocket to adjust its orbital parameters, allowing it, for example, to draw closer to the lunar surface for more detailed photographic coverage.[49]

Eastman Kodak, the pioneering giant of American photography, developed a special camera for Lunar Orbiter that enabled it to take blur-free images as it flew around the moon at thousands of miles an hour. A complex system monitored the spacecraft's altitude and velocity above the lunar surface, adjusting the film as each picture was taken to compensate for the camera's motion. Kodak's system produced photos of extraordinary quality, with resolution down to three feet, a vital aspect in helping select

Apollo landing sites.[50] The pictures were so good that the Soviet Union used them to map out its own potential manned lunar landing sites beginning in 1968, since they were better than any captured by their own spacecraft. Some NASA officials objected to making the Ranger, Surveyor, and Lunar Orbiter images public, fearing that the details could help the Soviet Union fly to the moon. But they were overruled because the 1958 legislation creating NASA mandated the "widest practical and appropriate dissemination of information . . . for the benefit of all mankind."[51]

Even as NASA's vast and successful program of scouting the moon was concluding, the Soviet Union continued with its own unmanned lunar missions under two programs, Luna and the later Zond series. Beginning with Luna 1 in 1959, the Russians carried out 20 successful missions of their own, including a number of important firsts: first probe to hit the moon; first flyby and photographs of the far side of the moon; first soft landing; first lunar orbiter; and first probe to circle the moon and return to Earth. Several of these feats were repeated as many as four times. Like the American space probes, several Luna lander missions were tasked with obtaining close-up photographs of the moon's surface to help assess the feasibility of manned lunar landings.[52]

Both sides also moved aggressively to send robotic space vehicles beyond the moon, to explore Venus and Mars, Earth's two neighboring planets. Both have fascinated mankind for centuries. Mars, with its mysterious surface "canals" fueling endless speculation about the possibility of life there, was the first planetary target of the space age. The Soviet Union launched two unsuccessful Mars probes in 1960, and three more, one intended as a Mars lander, were attempted in 1962. Though its communications failed en route, one of these, Mars 1, was the first spacecraft to fly past the red planet, at a distance of more than 122,000 miles, in June 1963. Two NASA Mariner spacecraft launched in 1964 to fly past Mars yielded a major success in July 1965, when one of them, Mariner 4, passed within 6,118 miles of the planet, returning 22 photos showing Mars' cratered surface and confirming the composition of its thin atmosphere. Mariner 3 failed, but the overall mission nonetheless was considered a great success in terms of increasing human knowledge of one of the Earth's closest planetary neighbors. Mariner 6 and Mariner

7 continued America's Martian exploration successes when both flew by the planet in 1969, sending home hundreds more surface photographs and additional surface and atmospheric temperature and pressure information. Both sides have continued sending spacecraft toward Mars, though the Russians have had very limited success despite their many attempts. Only one of its probes, Mars 5, successfully orbited and photographed the planet, in 1974.[53]

The Russians were more successful in their effort to explore Venus, even though the first successful mission there—actually the first machine from Earth to reach another planet—was the American Mariner 2 spacecraft. It flew past Venus in December 1962 at a distance of 21,600 miles, returning information on the planet's temperature and the thick clouds that covered it. However, the Soviets achieved a major coup in March 1966, when the Venera ("Venus") 3 spacecraft, a product of Korolev's design bureau, crashed on the Venusian surface. It was the first space probe ever to hit another planetary body. Venera 2, launched four days before Venera 3, flew past the cloud-shrouded planetary hot house at a distance of 15,000 miles. In 1970, Venera 7 became the first probe to send back data from the surface of another planet, an especially impressive achievement considering that Venus's surface temperature is roughly 932°F. The Russians' remarkable successes with Venus probes continued unabated into the 1970s and well beyond, including the first photos sent from the planet's surface in 1975 and later orbital missions to map the planet. The United States also continued to explore Venus.[54]

Only a handful of planetary probes, all of them American, have ventured beyond Venus, Mars, and Mercury. Given the vastness of space, trips to the outer planets—Jupiter, Saturn, Uranus, Neptune, and Pluto—require space probes able to survive a journey of years. The first of this hearty genre was Pioneer 10, launched in 1972. It flew past Jupiter in December 1973 to become the first spacecraft to reach the outer planets, going on to achieve another first as well when it left the solar system. In December 1974, Pioneer 11 also flew past Jupiter. Both probes provided color photos of Jupiter's clouds. Continuing its astonishing space journey, Pioneer 11 passed Saturn nearly five years later, in September 1979, providing information on the ringed planet's environment.[55]

Closer to home, Earth satellites continued to revolutionize life on a global scale, providing highly accurate weather forecasts and radio and television links. Other satellites mapped the Earth's surface and spied on the Soviet Union, monitoring its military activities. Of all these spacecraft, communications satellites have had perhaps the greatest significance for the lives of average Americans.

Arthur C. Clarke, noted author and scientist, predicted a future where artificial Earth satellites would link the globe electronically, a prophecy he made in 1945. He advocated placing three satellites in geosynchronous orbit. (Such satellites are placed directly above the Equator at a height of 22,240 miles. They remain over the same spot on Earth, circling the planet once each day.) Three such satellites could "cover" the Earth's entire surface, Clarke posited, and therefore be able to relay communications globally. With the advent of the space age, implementation of Clarke's concept became possible, though not right away. The earliest communications satellites were much less ambitious, starting with the Air Force's 1958 orbiting of an entire Atlas missile carrying a tape recorder that broadcast a message from President Eisenhower. That was followed in 1960 by Echo, an orbiting mylar-covered balloon that relayed radio signals bounced off its surface from one point on Earth to another.

The next major advance in satellite telecommunications was Telstar, an experimental satellite launched into orbit by NASA for American Telephone and Telegraph (AT&T) in July 1962. Telstar successfully relayed television signals for the first time from Europe (France) to the United States, a feat not possible with traditional TV broadcast signals. These signals travel in a straight line rather than follow the curve of the Earth, with a range of only about 50 miles from the point of broadcast. Telstar made international television a reality by amplifying and retransmitting the television signals across the Atlantic. By 1965, Hughes Aircraft had developed and orbited "Early Bird," the world's first operational commercial communications satellite, or comsat, as it came to be known. Early Bird, just as Clarke had written two decades earlier, was placed in geosynchronous orbit. It provided the capacity to handle hundreds of telephone calls simultaneously, offering a more economic alternative to the undersea cables. There was, however, one interesting

Unobstructed view: Soviet cosmonaut Alexei Leonov
takes his dangerous first "walk" in space, 1965.

First American in space: Onlookers at Cape Canaveral view Alan Shepard's blast-off, May 1961.

OPPOSITE: *People's hero: Excited and proud Russians line Red Square
to welcome cosmonaut Gagarin, April 1961.*

Mercury astronaut John Glenn contemplates the world he'll soon view from space.

Valentina V. Tereshkova, first woman in space, and Yuri Gagarin.

NEXT PAGE: *A confetti blizzard obscures onlookers during New York's March 1962 tickertape parade for John Glenn.*

Six of the seven original Mercury astronauts mug for the camera, February 1961.

Welcome home: Yuri Gagarin acknowledges applauding soldiers
soon after return from orbit, April 1961.

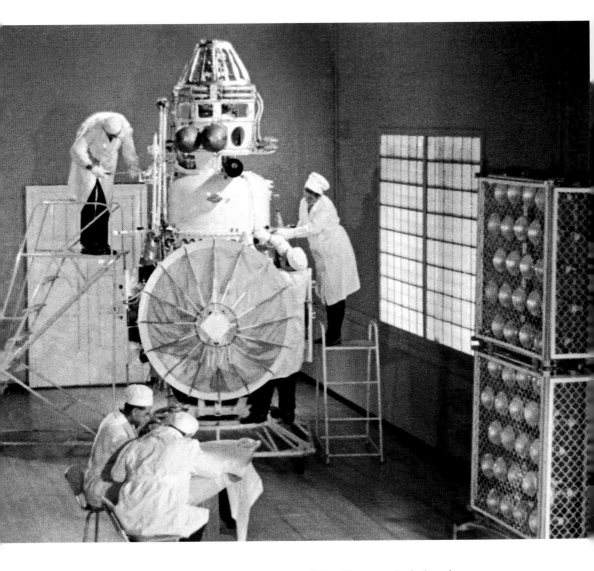

Soviet technicians prepare a 1969 Venera ("Venus") space probe for launch.

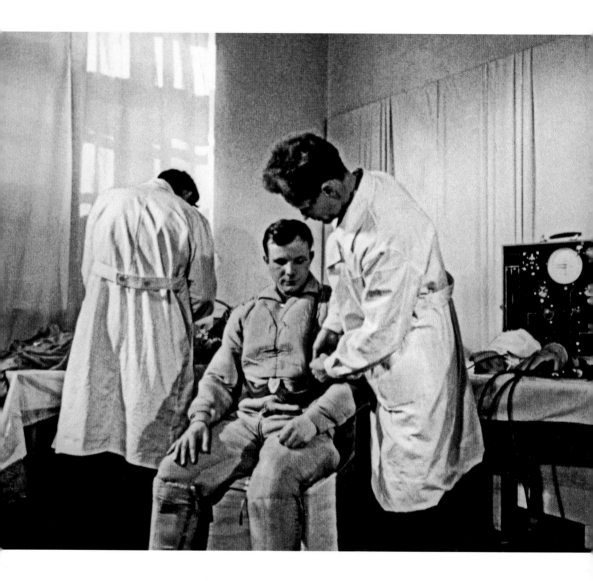

Gagarin undergoes medical tests during training for his orbital flight.

NEXT PAGE, LEFT: *Astronaut Buzz Aldrin during jungle survival training in Panama, 1964.*

NEXT PAGE, RIGHT: *Mercury astronaut Alan Shepard trains in simulator, 1959.*

Soviet cosmonaut Komarov (foreground) on cross-country skis.

A hero lost: Muscovites read about Gagarin's death, March 1968.

"The Chief Designer" passes: Moscow funeral procession for Sergei Korolev, January 1966.

Fellow astronauts accompany the body of Gus Grissom, killed in Apollo 1 fire, January 1967.

Apollo 1 astronauts Chaffee, White, and Grissom (from left) practice in Apollo mission simulator.

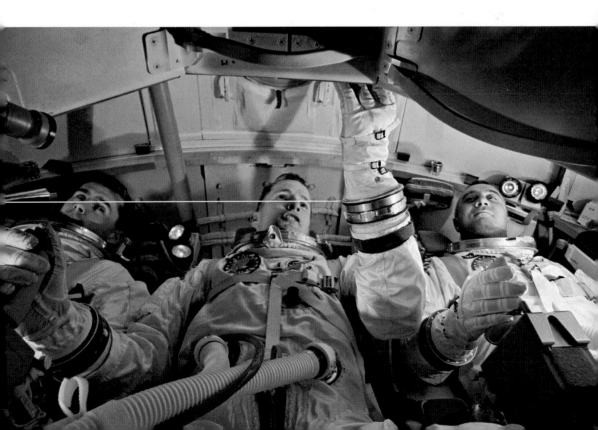

Astronauts in improvised desert-survival training clothing made from parachute material.

Von Braun (second row, second from right) and colleagues observe Saturn launch, May 1965.

ABOVE, TOP: *First American into space: Alan Shepard aboard Freedom 7, May 1961.*

ABOVE, BOTTOM: *Astronauts practice post-landing egress from Apollo spacecraft.*

OPPOSITE: *Saturn I rocket, precursor to the larger Saturn V, blasts-off from Cape Kennedy, January 1964.*

Sue Borman (front left) cheers husband Frank Borman's return from the moon, December 1968.

OPPOSITE: *Apollo 9 astronaut Dave Scott exits command module for spacewalk, March 1969.*

NEXT PAGE: *Astronauts briefly float gravity-free in an Air Force plane.*

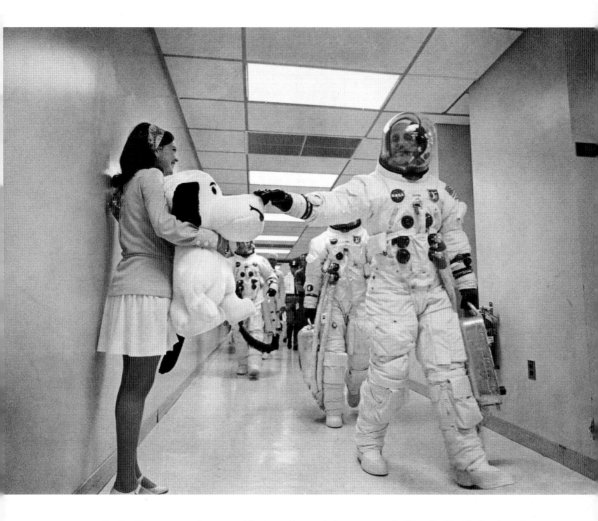

Apollo 10 mascot receives a good-luck pat from mission commander Tom Stafford before launch.

OPPOSITE: *Regaining his Earth legs: astronaut John Young exits the Apollo 10 command module, May 1969.*

FOLLOWING PAGE: *John Glenn relaxes on a Navy destroyer after his February 1962 orbital flight.*

exception to Clarke's original proposal: Signals from geosynchronous comsats do not transmit well to Earth's northern regions. This was an important issue for the Soviet Union, since much of its landmass lies in the far north. The Soviets solved this problem by developing their own system of Molniya ("lightning") comsats, which use a highly elliptical, or egg-shaped, orbit carrying them high over the northern hemisphere.[56]

From the mid-1960s comsats proliferated in number, capability, and sophistication. They permanently have altered human communication by voice and data transmission and have provided entertainment and news in the form of television and other electronic signals. Among all the changes that came with the space race, communications satellites have had a profound and continuing influence on modern life.

GETTING HOME

When asked to name the most beautiful thing he had ever seen on a space mission, astronaut Wally Schirra replied, "The parachute opening." A spacecraft in orbit has reached a kind of summit, an achievement won through great effort. However, like a mountaineer on the peak, a spacecraft in orbit has accomplished only half of the journey. "We established very clearly at the outset that this was supposed to be a round trip," Schirra used to say. Getting home, from a summit or from orbit, takes getting down, and the route back must be negotiated with care, for it can prove just as fatal as the route up.

To return to Earth from space, a spacecraft must line itself up on a precise approach trajectory to hit the Earth's atmosphere at just the right angle. Much like a jet lining up on a short runway in the Andes, the spacecraft has very little room for error. Too shallow an angle, and the spacecraft could shoot back out of the atmospheric curvature and fly off into space without the power to return. Too steep an angle, and the spacecraft will burn up. An Apollo spacecraft had to hit the atmosphere within half a degree of 6.49 degrees in order for it to begin a successful reentry run.

The main problem of reentry from orbit into the Earth's atmosphere is like the problem of getting there in the first place: It is not so much the height as it is the speed. If a spacecraft at orbital altitude could simply free-fall straight down to Earth like a skydiver, reentry wouldn't really be so difficult. The orbiting spacecraft cannot do this, however, because of the high speed at which it is moving. The spacecraft could fire braking rockets to slow it to a gentle speed—if it carried strong enough engines and sufficient fuel. But it took the strength of an entire rocket booster to accelerate the spacecraft from zero to orbital velocity of 17,500 miles per hour. The energy to cut that speed back down to zero has to come from somewhere. No space vehicle yet built has been able to bring that kind of rocket power with it into orbit, so rocket engineers use atmospheric friction to slow the returning spacecraft for a survivable landing.

Home from the moon: Apollo 8 photographed during Earth reentry, December 1968.

Air friction offers a considerable braking force, but friction causes heat, just as you can quickly heat your hands with a vigorous rub. At the "vigorous" speed of a supersonic airliner, the nose of the Concorde commonly reached 260°F, with the rest of the plane's fuselage topping out at around 200°F, almost hot enough to boil water. The Concorde encountered this boiling heat at Mach 2. From an orbital velocity of Mach 25, a spacecraft dropping into the atmosphere must endure air-friction temperatures in the thousands of degrees. Constructing spacecraft

to survive this harrowing trial challenged both Soviet and American engineers, and while the basic problems were the same, each side worked out slightly different solutions.

American space capsules took the form of squat cones with a rounded blunt end built as a heat shield designed to take the brunt of reentry punishment. The Apollo moon missions raised the challenge even higher than orbital reentry, increasing spacecraft speeds up to Mach 32, or some 25,000 miles per hour. The capsule and its returning astronauts would be surrounded by temperatures reaching

5,000°F, as hot as the sun's corona. Nothing that we could engineer could long endure the full force of such an assault, so Apollo engineers built the spacecraft's heat shield on the principle of *ablative* material: Phenolic resin in the cells of a honeycomb shield was designed to melt, absorb the tremendous heat, and then flake away, taking the heat with it.

To reduce the force of the reentry, Apollo spacecraft used small maneuvering thrusters to afford it some steering control through its orientation. Blunt face on, the capsule produced maximum drag and fell. Designed with the center of gravity offset, the ship would tilt forward slightly, and in this orientation the capsule produced some lift, and could "surf" upward for a limited time. With these techniques an Apollo capsule traced a roller-coaster path, plunging the capsule into the thick of the atmosphere to burn off speed, but then tilting it upward to rise and cool before turning down again for another plunge. Two such peaks cut down on the maximum heat levels and reduced the required thickness of the heat shield.

The roller-coaster Apollo reentry trajectory also helped reduce the *g*-forces that had to be endured by the astronauts aboard. Slamming on the brakes in your car while traveling at highway speed throws you forward with some force. Imagine hitting the brakes at 25,000 mph! The astronauts, lying on their couches with their backs to the heat shield, were hit with a growing "brake-effect" deceleration force that at its peak of 6.35 *g*s squashed each astronaut's body with the weight of over a thousand pounds.

The Apollo space capsule design afforded maximum control through high technology and a complex design, a typically American profile. Soviet designs show a contrasting emphasis on economy and simplicity. Soviet Vostok and Voskhod space capsules were spherical and had no reentry maneuvering controls. They relied on "passive" attitude control during reentry: The positioning of heavy equipment inside a capsule put the craft's center of gravity deliberately off-center, and thus made the sphere naturally turn one face into reentry. Uncontrolled, the capsules had to plunge like stones in a simple "ballistic" reentry path, which built up a maximum force of over eight *g*s. Fighter pilots typically black out at nine.

American space capsules were single-room units, little more than flying cockpits until the debut of the Apollo capsule, which offered some additional usable interior space. In all of these spacecraft, the entire capsule was constructed to survive reentry. Soviet spacecraft designers took a different approach with the Soyuz craft that became the mainstay vehicle of cosmonaut missions. The Soyuz included a "living room" orbital module in addition to the "cockpit" reentry module, offering two distinct spaces in the vehicle. Gear that did not need to be brought home such as dining, lavatory, and orbital operations equipment was all located in the orbital module, which was closed off and discarded before reentry. The remaining reentry module needed only to contain the cosmonauts themselves, so it could be made small and compact, minimizing the amount of heavy (and expensive) construction that must be built to withstand reentry forces, another typically economical Soviet choice.

Reentry subjects the delicate components of a spacecraft's complex operating system to tremendous strains and has proven to be the most dangerous portion of a space traveler's journey. On the Soyuz 1 mission in 1967, Vladimir Komarov survived a harrowing series of orbital malfunctions that forced him to make a dangerous manual retrofire to begin his reentry. After Komarov's capsule emerged successfully from the hot phase of the return, in a heartbreak-ing twist of fate, his parachutes failed. Trapped in the free-falling capsule, Komarov died upon impact.

Soyuz 11 made a soft landing in 1971, but recovery crews opened the hatch to find that all three cosmo-nauts—who could not wear space-suits in the cramped capsule—were dead. Just before reentry, a faulty valve had opened and bled all their oxygen into space, sucking the very breath from their lungs. The auto-mated craft had returned their life-less bodies to the Earth.

It is only in these tragic failures that we see clearly the specter haunt-ing every return to Earth. Through them we appreciate the extraordi-nary accomplishments of the design-ers who make successful returns pos-sible. The dire prospect of a Komarov fall, a spacecraft's disintegration, or any other mode of fatal failure drives engineers to build into each vessel the very best defenses they can create.

Each space traveler who boards a spacecraft for launch prepares to enter the realm of the heavens and the pages of history, to see sights and dare deeds reserved for a privileged few—but as they strap in and as the hatch is closed behind them, they must anticipate these rewards in the knowledge that the fires of reentry may exact the ultimate price. ■

7

A DISTANT PRIZE

By the mid-1960s, Project Apollo was moving in full stride toward the goal of a lunar landing. The Apollo missions were the culmination of America's building-block approach to manned spaceflight. Project Mercury had provided the first manned flights, ranging from the early up-and-down suborbital ventures to short-duration orbital flights, all for single astronauts. The next vital step was Gemini, a much more sophisticated undertaking that launched two astronauts on orbital journeys lasting as long as two weeks. Gemini enabled the United States to develop skills essential to going to the moon, including orbital rendezvous and docking techniques. The American space program was now poised to pursue its most ambitious goal, the final lap to the moon. NASA allocated massive financial and human resources to achieve this end. Apollo's success rested on the ability of NASA to mobilize a vast cadre of talented managers, engineers, scientists, and astronauts to achieve this singular goal. There remained the all-consuming technical challenge of

OPPOSITE: *Neil Armstrong's reflection is visible in this iconic photo of Buzz Aldrin on the moon, July 1969.*

designing, testing, and flying spacecraft capable of reaching the distant prize of the moon.

NASA was not alone in this endeavor. The Soviet Union had demonstrated in the past a formidable capacity to pull off space spectaculars; many NASA observers regarded this technical prowess as preparation for a Soviet manned lunar mission in the near term. They also felt the pressure of fulfilling the mandate of President John F. Kennedy, expressed on May 25, 1961, to send humans to the moon by the end of the 1960s.

From the start, Kennedy's call for a lunar landing had resonated with the American public at large. Congress had expressed its support through lavish financial backing. These factors had helped to instill in NASA a keen sense of institutional purpose and high morale. The stakes were high. The whole enterprise was pursued in the Cold War context, which meant that anything less than triumph would lead to national humiliation. Going back to the early Kennedy years, then Vice President Lyndon Johnson and other high officials had examined the question of manned spaceflight and concluded that the United States had a "reasonable" chance to overtake the Soviet Union.[1]

Nevertheless, the managerial tasks faced by NASA head James Webb remained daunting. "The Apollo requirement," Webb observed, "was to take off from a point on the surface of the Earth that was traveling at 1,000 miles per hour as the Earth rotated, to go into orbit at 18,000 mph, to speed up at the proper time to 25,000 mph, to travel to a body in space some 240,000 miles distant, which was itself traveling 2,000 mph relative to the Earth, to go into orbit around this body, and to drop a specialized landing vehicle to its surface." Moreover, he concluded: "There men were to make observations and measurements, collect specimens . . . and then repeat much of the outward-bound process to get back home. . . . One such expedition would not do the job. NASA had to develop a reliable system capable of doing this time after time."[2]

As NASA administrator, Webb would be at the epicenter of this project during much of the 1960s. His leadership played a large role in the ultimate success of Apollo. Before coming to NASA in 1961, Webb had carved out an impressive career as a capable bureaucrat and Washington insider. He had served as director of the Bureau of the Budget and undersecretary of state during the Truman administration. He was adept at negotiations with Congress and various entities in the federal government. He built a broad and powerful constituency for NASA. His tenure oversaw the construction of new facilities, mainly in the South, and the award of massive contracts to the aerospace industry to build the essential spacecraft and components for the staged NASA space program.[3]

NASA gave considerable attention to the development of spacecraft for the manned spaceflights. As early as mid-July 1961, Webb invited hundreds of aerospace industry representatives to attend a NASA-Apollo Technical Conference in Washington. At the conclave, each representative received written guidelines containing the technical specifications for a future Apollo spacecraft, including a command module, a service module, and a lunar landing module. Each component was deemed essential to transport humans from Earth to the lunar surface and back. The number of technical details for such an ambitious project was mirrored in the size of the written guidelines, which weighed a gargantuan 250 pounds.[4]

One key debate within NASA focused on the optimal method of landing on the moon. This "choice of mode" was a pivotal concern, one that shaped ultimately the design for the launch rocket and the spacecraft. Early on, there was great uncertainty over Soviet intentions and capabilities. Moreover, the engineering challenges were profound and unprecedented: The NASA leadership was fully aware of the fact that going to the moon would be a pioneering journey, and there was no guarantee of success.

One option, later to be dubbed the "Direct Ascent" approach, was advanced by von Braun and his Huntsville team. The proposed technique for a lunar landing was simple and straightforward: A rocket fired from Earth took a trajectory that would "lead" the moon—in the same fashion as a hunter aims in front of a flock of geese—so that the two objects arrived at the same place at the same time. But this approach had significant drawbacks. It would require the development of a huge rocket called Nova, which von Braun envisioned with 10 engines in the first stage alone. Such an immense space vehicle—with a flyable stage for the return flight to Earth—would have been complex and costly to build.[5]

A second option was Earth orbit rendezvous (EOR). This plan called for the assembly in orbit of separately launched components—the spacecraft, lunar lander, and equipment module. The EOR envisioned a sequence of launches, which promised less expenditure of energy to get a lunar mission under way. Writer William Burrows aptly described this option as a "high-tech mule train" to the moon. The detractors of the EOR pointed out its cost, difficult assembly, and inherent dangers. Such a complex enterprise struck many engineers as impractical, even foolhardy.[6]

A third scheme finally prevailed. Known as lunar orbital rendezvous (LOR), it had to overcome strong initial resistance.[7] The plan was advocated by John Houbolt, a structures engineer at NASA's Langley Center, who advocated his ideas with great enthusiasm. The LOR required that the Apollo command and service module (CSM), be placed in a "parking" orbit around the moon, while the lunar lander carried astronauts down to the lunar surface. Initially, fierce opposition greeted the idea, both at Langley and NASA headquarters. Concern was voiced that if the astronauts were unable to rendezvous with the CSM when they left the moon, they would almost certainly die in lunar orbit, their lander becoming their crypt. Despite the opposition, Houbolt refused to give up. In November 1961, bypassing many management levels, he wrote an impassioned letter advocating LOR directly to NASA associate administrator Robert Seamans, Jr. In it, he characterized himself as "a voice in the wilderness." His perseverance won him a serious review of his proposal, though it took considerable time and effort to win the doubters over to his side. Houbolt, who had joined NASA's predecessor agency, NACA, in 1942, received the prestigious NASA Medal for Exceptional Scientific Achievement for his LOR work.[8]

What ultimately carried the day for LOR was the fact that it required fewer resources than either of the other two options. Moreover, the rival Direct Ascent and EOR modes would have required a massive rocket— perhaps as tall as six stories—to land on the moon. By contrast, the LOR mode required only a single Saturn V, much smaller than the proposed

Nova "super booster" for Direct Ascent, for example. In addition, using an orbiting command module to retrieve the astronauts as they ascended from the moon meant the weight and design of the lunar lander could be simpler, smaller, and lighter. That was because part of the lander would remain on the moon when its "ascent module" blasted off to carry the crew up to meet the command module. Von Braun eventually endorsed the LOR mode because it alone offered a solid chance to meet Kennedy's lunar goal on schedule. NASA publicly announced selection of the LOR mode on July 11, 1962.[9]

NASA grew exponentially during the 1960s. Employment rolls increased to 36,000 in 1966, a significant leap from the 1960 level of 10,000 employees. NASA's management had decided at the start of the manned lunar program that it would be most efficient to rely on outside researchers, universities, and private industry for the bulk of the work. As a result, the total number of persons in these three categories working on Apollo expanded from 36,500 in 1960 to an astonishing 376,700 in 1965. Project Apollo required federal funding for NASA commensurate with the program's lofty goals, which Congress provided. NASA's budget exploded from 500 million dollars in 1960 to 5.2 billion dollars in 1965, representing 5.3 percent of the entire federal budget that year. The following year, with Apollo's costs largely funded and the increasing budgetary demands of the Vietnam War, NASA's budget began a decade-long decline.[10]

A considerable portion of Apollo expenditures was allocated for three new facilities. One was the Manned Spacecraft Center in Houston, later renamed the Lyndon B. Johnson Space Center, responsible for developing the Apollo spacecraft in addition to its role in training the astronauts and serving as Mission Control. Ground testing of the huge Saturn rockets for Apollo required the construction of the Mississippi Test Facility (later renamed the John C. Stennis Space Center) on a large bayou. Finally, NASA developed the immense Merritt Island spaceport adjacent to the Air Force's Cape Canaveral launch pads, including pads for the moon rocket and the Vehicle Assembly Building to stack its components.[11]

Webb and his deputies coordinated the many NASA program offices in Washington and in the field. Cordial relations with the White House and Congress were paramount, taking a huge amount of time and energy. They also had to maintain relations with other federal agencies and the scientific community, both of which did not always agree with NASA's policies. Separate from the government's largely Washington, D.C.–based constellation of participants were Apollo's prime contractors and their thousands of subcontractors. It was imperative that these participants work efficiently with NASA and with each other. The most visible aspects of Apollo—the moon rockets and the astronauts—were thus only part of a complex effort.[12]

Webb and his senior staff sought to keep Apollo on schedule. In the fall of 1963, George Mueller, newly appointed as NASA's associate administrator for manned spaceflight, asked his staff to estimate the chances of Apollo meeting its decade-end goal. After surveying various NASA programs, including the troubled and significantly behind-schedule Ranger unmanned lunar exploration project, they reported that the odds were only one in ten. Mueller, with a Ph.D. in physics and years of private-sector experience in ballistic missile development, realized that new approaches were needed to avoid significant delays in the program. His solution, promulgated in a November 1963 memorandum, was to shorten the entire Saturn test program by rejecting incremental testing for a concurrent or "all-up" approach. This meant that the mighty Saturn V moon rocket would fly with all three of its stages "live" on its first flight, rather than each stage being tested and flight-qualified individually.[13]

At first Von Braun and his team resisted when Mueller came to Huntsville in 1964 to introduce his new test philosophy. "To the conservative breed of old rocketeers, who had learned the hard way that it never seems to pay to introduce more than one major change between test flights, George's idea had an unrealistic ring," von Braun wrote later. Compared with testing the Saturn rocket 'a stage at a time,' "adding a second stage only after the first stage had proven its flightworthiness, [Mueller's] 'all-up' concept was startling," von Braun noted. "It sounded reckless, but George Mueller's reasoning was impeccable. . . . In retrospect it is clear that without all-up testing, the first manned lunar landing could not have taken place as early as 1969."[14]

As part of his effort to keep Apollo moving, Webb also recruited Air Force Major General Samuel Phillips, a World War II combat veteran who later directed the Air Force's highly successful Minuteman ICBM program. Minuteman, the nation's third ICBM, entered service on time and under its projected cost. Joining NASA in 1962, Phillips became Project Apollo director in December 1964. He brought with him dozens of other Air Force officers experienced in the management skills and procedures needed for fast-track ballistic missile development. They successfully applied these procedures to Apollo, including centralized authority over design, engineering, logistics, and other critical program aspects and the use of formal management reviews, to keep the program on schedule. Phillips' management operation efficiently oversaw and coordinated the work of more than 500 Apollo contractors as well as NASA's own major contributions to the lunar effort.[15]

PERILS OF THE SPACE AGE

The year 1967 set the stage for dramatic reversals in the space programs of the United States and the Soviet Union—unwelcome and grim reminders of the high risks associated with space technologies. At dusk on the evening of January 27, the crew of the Apollo 1—Virgil "Gus" Grissom, Edward White, and Roger Chaffee—began a series of pre-launch simulation tests high atop their Saturn IB on Pad 34 at Cape Kennedy. The projected Apollo mission, scheduled for a February launch, would signal a new phase in America's drive to make a landing on the moon. The newly designed Apollo command module accommodated a crew of three astronauts. On this evening they were supposed to conduct the vital "plugs out" test, where all electrical and ground links were disconnected to verify that the spacecraft could function solely on internal power. Nearby a large cadre of technicians assisted with the test, scattered among the launch facility in a blockhouse, the service structure, the swing arm of the umbilical tower, and the Manned Spacecraft Operations Building. The simulation tests were monitored from a nearby concrete bunker.[16]

The Apollo 1 crew was led by the veteran "Gus" Grissom, a member of the original Mercury 7. He had flown on the second sub-orbital mission in 1961, that ill-fated flight where the Liberty 7 capsule had been lost in the Atlantic after the hatch door had blown away prematurely and Grissom had nearly drowned. In 1965, he flew on a successful Gemini mission, adding to his impressive flight log as an active astronaut. A combat veteran, he had flown as an Air Force fighter pilot in the Korean War. Now 40 years old, Grissom enjoyed status as an experienced member of the elite astronaut corps, a man highly regarded for his personal dedication and technical competence.

Edward White, also an Air Force pilot, had won fame on the Gemini 4 mission when he became the first American to walk in space. A San Antonio native, White was a West Point graduate and a former Air Force test pilot. Roger Chaffee, a Navy lieutenant commander, came to the Apollo program with a bachelor's degree in aeronautical engineering from Purdue University. Apollo 1 was to have been Chaffee's first spaceflight. The Grand Rapids, Michigan, native was a member of NASA's third astronaut class, selected in October 1963, and had logged more than 2,300 hours of flying time.

When the three astronauts entered the command module, they began their simulation sequence in a pressurized chamber of 16.7 pounds per square inch of pure oxygen. As the crew went through the detailed checklist, they encountered several communications problems, which prompted delays. Links to the outside were maintained on four channels, by radio or telephone. Searchlights illuminated the Saturn booster rocket, which towered 20 stories above the launch pad. The crew had performed the same routine in the simulator, and the Saturn rocket was not fueled. All seemed routine until a garbled message suddenly came from the Apollo command module—"Fire!" A second message quickly followed, sent by Roger Chaffee: "We've got a fire in the cockpit!" Television monitors revealed Ed White's arms reaching back to release the bolts that held the hatch shut. A television camera caught an ominous yellow glow from the interior of the capsule during these critical seconds. Flames soon appeared at the hatch window. In this enveloping maelstrom of fire and smoke, pad technicians looked on in horror as silver-clad arms attempted in vain to open the hatch. As the seconds passed, a third message reached the controllers: "We've got a bad fire. . . . We're burning up!" An emergency crew rushed to the capsule and attempted to open the hatch door, only to be driven back by the intense heat. The temperature inside the cabin reached 2,500°F. Repeated efforts to reestablish communications with the astronauts failed.

Silence reigned when the hatch was finally opened. Two NASA physicians hurried from the blockhouse control center to the launch pad, only to discover that nothing could be done to save the lives of the trapped astronauts. Subsequent autopsies revealed that the three men had died from asphyxiation by the inhalation of toxic gases, not from their severe burns. Their nylon suits had melted and fused together in the intense heat of the fire. The tragedy cast a pall over the NASA space program. Three astronauts—Elliot See, Charles Bassett, and Theodore Freeman—had died in airplane crashes, but over the course of six years, no astronaut had perished in any space launch procedure.[17]

The Apollo 1 fire literally shut down the American space program for months. Webb set up a fact-finding committee under Floyd L. Thompson, the head of the Langley Research Center. Congress took a keen interest in the affair, as did the media. The NASA leadership feared that the Apollo program could be severely delayed or even cancelled. Congress held hearings to assess the reasons for the tragedy. Early in April, the Thompson committee made its report: a detailed set of findings divided into 14 booklets and numbering more than 3,000 pages. While the report did not cite one specific reason for the fire, it did identify a series of factors that led to the tragedy, among them: the pressurized cabin filled with pure oxygen; the existence of combustible materials; faulty wiring; and inadequate means for egress from the capsule. NASA had made the decision to use pure oxygen for simplicity and weight considerations. Notwithstanding all the safety measures, a spark had ignited the fire within the capsule, later traced to defective wiring in the lower equipment bay at the foot of Grissom's seat. The Thompson committee findings implied that NASA and its contractors had failed to provide proper design, engineering, and quality controls for the Apollo spacecraft. Debate would rage over the accident in the years that followed, and some critics of NASA asserted that the space agency had sacrificed safety in the intense drive to overtake the Russians in the space race. Webb found himself in the crosshairs of public criticism throughout the Apollo 1 crisis, which placed enormous pressures on him. He left NASA in October 1968. His departure coincided with the final phase of the Apollo program.[18]

The Soviet space program faced its own setbacks during this formative period, although knowledge of these reversals was not fully disclosed until years later. As noted earlier, on March 23, 1961, cosmonaut trainee Valentin Bondarenko perished in a flash fire in a simulation test. He and other trainees spent many hours, even days, in an oxygen-saturated chamber. The 23-year-old Bondarenko died when he attempted to change sensors on his body, using an alcohol cotton swab. He dropped the swab on a red-hot spiral of an electrical element used for heating food. The resulting fire enveloped the chamber and killed Bondarenko. If the Soviets had been more open about the tragic fire and the dangers associated with a pure-oxygen environment, Bondarenko's fate could have served as a cautionary tale for NASA contractors.[19] Soviet cosmonaut Aleksei Leonov thought they did know. He followed the Apollo 1 story with great interest. "The Americans must have known of the tragedy that had befallen Bondarenko in a pure-oxygen environment," Leonov argued many years later. For him, the "American intelligence services would not have been doing their job properly if they had not informed NASA about what had happened." For whatever reasons, the lessons to be learned from the Bondarenko incident never influenced the NASA program in any discernable way.[20]

The Soviets also experienced disaster in that fateful year. On April 24, 1967, just months after the Apollo fire, Vladimir Komarov died during an aborted mission of the newly designed Soyuz 1 spacecraft. The Soviets had planned for Komarov, a talented test pilot, engineer, and veteran of the Voskhod series, to be the pathfinder on a remarkable space spectacular. The script called for Komarov to fly solo into orbit and then be joined the following day by another Soyuz spacecraft with three cosmonauts aboard. The two orbiting space vehicles would rendezvous and then dock, setting the stage for two cosmonauts to transfer to Komarov's Soyuz 1 in an elaborate, high-risk spacewalk maneuver. Once the transfer had been completed, the second Soyuz spacecraft would return that same day with one cosmonaut on board. This maneuver would establish a new benchmark even as it perfected skills deemed essential for any future lunar mission.

The well-laid plans for the space docking never materialized. No sooner had Komarov reached orbit than his Soyuz 1 began to develop problems. First, one of two solar panels failed to deploy properly. This loss of a solar panel meant a dramatic reduction in power for the guidance and other critical systems of the spacecraft. Soon Komarov faced serious problems with attitude control of his spacecraft. In response, Soviet mission controllers called for Komarov to return as soon as possible and cancelled the second Soyuz launch. At reentry, Soyuz 1 entered the upper atmosphere at high speed, and the spacecraft spun out of control. The Soyuz 1's parachutes failed to deploy properly, becoming entangled, and the spacecraft was propelled in a high-speed downward trajectory. Komarov was killed instantly. The rescue team found the crashed spacecraft near Orsk, at the tip of the Ural Mountains near the border of Kazakhstan. In the aftermath, the inquiries into the cause of the accident discovered that the complex design for the parachutes had not functioned properly, just one defect in a cluster of design and engineering problems associated with the new Soyuz design. Komarov had gone aloft in an untested spacecraft. Its essential design was sound, and in the years that followed, it evolved into a reliable spacecraft, one that would remain at the epicenter of the Soviet space missions. In the context of 1967, however, the tragic accident dealt a powerful blow to the Soviets, similar to the consequences of the Apollo 1 accident.[21]

WERNHER VON BRAUN AND THE SATURN ROCKET

As the American space program unfolded, Wernher von Braun assumed an important role as the prime architect of the Saturn rockets. At this juncture he occupied a key position in NASA planning. Earlier, in January 1958, he and his innovative team of German rocket scientists had lofted America's first satellite, Explorer 1, redeeming Vanguard's very public launch-pad failure in December 1957. At the time of this triumph, von Braun already was thinking and planning for the time when the United States would send astronauts, space probes, and other very heavy payloads on adventurous journeys into Earth orbit and far beyond. In the spring of 1957, months before Sputnik 1 was orbited, von Braun and his Huntsville team had begun design studies for a new, massive multi-engine rocket booster, capable of lofting payloads far heavier than any existing launch vehicle. Given further impetus by the October launch of Sputnik, von Braun's proposed super-booster continued to evolve and received funding for development in August 1958.[22]

The appeal of von Braun's proposal, known as Saturn I, lay in its very concept: It would utilize existing components as building blocks to create the brute lifting power for a new generation of high-payload launchers. The first stage of Saturn I consisted of eight Redstone boosters, each with an H-1 rocket engine, an upgraded version of the engine used on the Army's Jupiter missile. Together, the engines produced 1.5 million pounds of thrust. The engines and the tankage for their liquid oxygen (LOX) and kerosene fuel were clustered around a central core adapted from the Jupiter missile. Von Braun and others jokingly referred to the Saturn I first stage as "Cluster's last stand."[23] Saturn I's second stage was powered by four engines fueled by LOX and liquid hydrogen, a mix providing greater thrust than LOX and kerosene. However, this mixture was highly volatile, required special handling, and proved very difficult to develop.[24]

When they began work on Saturn I, von Braun and his team were still part of the Army Ballistic Missile Agency (ABMA). The Saturn program was transferred to NASA late in 1959 and the ABMA followed in 1960, despite fierce resistance by the Army to giving up its space program. The Army's missile facility in Huntsville, Alabama, was renamed as NASA's George C. Marshall Space Flight Center, and von Braun became its director. Throughout this period, which at times caused his loyal team considerable uncertainty about their future, he did not waver in his reassurances. He told them he had no doubt about an eventual "moon-landing project" that would include a major role for "our team."[25] Events would prove his statements true: The start of the Saturn development was a major milestone en route to the moon. While an important figure in the NASA organization, though, von Braun was not solely responsible for the success of the manned lunar program. He was part of a team. The combined creativity, managerial skills, brilliance and foresight of people like Webb, Gilruth, Phillips, Houbolt, Dryden, and others were needed to drive the American space program forward in the 1960s.

Saturn I, a concept demonstrator for multi-engine heavy-lift rockets, carried out 10 successful flights beginning in October 1961. Its successor, the Saturn IB, was upgraded with more powerful first- and second-stage engines. In January 1968, it boosted the Apollo 5 mission, placing the lunar module and an unmanned prototype lunar lander into Earth orbit.[26]

Saturn I and IB essentially were the warm-up acts for the main event: Saturn V, the rocket that would deliver humans safely to the lunar surface. Saturn V was an astonishing and elegant engineering triumph, unprecedented by any measure. It was 364 feet tall; fully loaded, it weighed 5.8 million pounds; and a total of 11 engines powered its three stages, producing 8.7 million pounds of thrust. The first stage was powered by five F-1 kerosene and LOX-fueled engines, each with a thrust of 1.5 million pounds. The second stage contained five J-2 engines running on liquid oxygen and liquid hydrogen, and the third stage used a single J-2 engine. Saturn V's mighty engines could place more than 124 tons in Earth orbit or propel 50 tons toward the moon at 25,000 miles per hour.[27] The launch vehicle generated superlatives at every turn, requiring a 52-story building at the Cape just to assemble it in a vertical position before it was rolled, complete, to its launch pad. Some have compared Saturn V to the construction of the Great Pyramid in ancient Egypt, noting that the Greek historian Herodotus had been told during a visit to the ancient site in 450 B.C. that 400,000 people had labored to build it. In the 1960s, nearly 2,500 years later, NASA employed a similar number of people on Project Apollo, with Saturn V the most visible product of their work.[28]

Saturn V's first stage was so huge that once assembled it had to be transported by barge through the Gulf of Mexico and up the Pearl River to NASA's isolated Mississippi Test Facility (later the Stennis Test Center). Only then, mounted on a special test stand, was the first stage far enough away from civilization to allow full runs of its engines without causing damage. Early tests of the F-1 at the Marshall Space Flight Center in Huntsville had caused windows to break and dishes to fall off walls miles away.[29]

As spectacular as it was, Saturn V was only part of the massive undertaking. The lunar journey would require parallel development of a unique spacecraft to carry humans to the moon, allow them to land and explore it, and then to return home safely. The Apollo spacecraft was composed of three parts stacked on and inside the Saturn V: the command and service modules, attached to each other and set atop the huge rocket, and, behind them, the lunar module, housed in a garage-like section of the rocket.

The command module (CM) contained the three-man crew's quarters and flight controls, and the service module (SM) housed the propulsion and spacecraft support systems. Joined together, these two modules were referred to as the CSM. The lunar module (LM) was designed to carry two astronauts to the lunar surface, support them while there, and then return them to the CSM in lunar orbit.

Of the three components, each an extraordinary feat of engineering, one by definition stood out: the lunar module, which would carry humans to another celestial body for the first time. Since it would operate only in the vacuum of space, it did not have to be sleekly designed like an aircraft. In fact, it looked quite ungainly, a collection of odd angles mounted on landing struts, resembling a large metallic insect. Anthropomorphically, its twin triangular windows could be seen as eyes, and the centered main hatch as a mouth. Completing the portrait were more than a half-dozen antennae atop the LM.

Its unusual look sprang from the unique missions, which greatly influenced its design. It consisted of two sections. One was the descent stage, holding the rocket engine used for the descent and the landing gear; this stage would be left on the lunar surface. Set on top of the boxy descent stage was the ascent stage, consisting of the crew cabin, with its instrument-packed control panel, supplies, oxygen for breathing, and a separate rocket to blast the ascent stage back into lunar orbit for its rendezvous with the CSM.[30] The driving force in the LM's design was to make it as lightweight as possible. To achieve that, any aerodynamic curves were eliminated, along with seats for the crew, unnecessary in the weightlessness of space or on the moon, which has only one-sixth of Earth's gravity. The imperative to eliminate every possible ounce resulted in the descent stage being skinned only with Mylar wrapping stretched over a frame. The crew cabin walls of the ascent stage were as thin as five-thousands of an inch; a screwdriver accidentally dropped by a worker inside the cabin during testing fell right through the floor.[31]

President Kennedy visited Cape Canaveral on November 16, 1963, just six days before his assassination in Dallas, Texas. He had come to inspect the Saturn I launch pad and to be briefed by von Braun on its capabilities. He also flew by helicopter over nearby Merritt Island, where construction of NASA's huge Saturn V launch facilities was then under way. Kennedy showed great interest in a set of scale models illustrating how the Mercury-Atlas rocket that had carried the first American into orbit was completely dwarfed by the Saturn V. Kennedy's timely visit stirred great enthusiasm among the personnel working on the NASA space program. Following Kennedy's tragic death, his newly sworn-in successor, Lyndon B. Johnson, renamed both the NASA and Air Force launch facilities located at the Cape as the new John F. Kennedy Space Center and Cape Kennedy Air Force Station, respectively.[32]

THE SOVIET SPACE PROGRAM IN CRISIS

American intelligence experts took a close look at the Soviet space program in 1965. Their report, summarized in the National Intelligence Estimate (NIE) for that year, focused on the Soviet Union's "capabilities and probable accomplishments in space over the next five to ten years." The NIE noted, "The Soviet space program will retain its priority, that its accomplishment will continue to be impressive, and that it will focus on goals for which the U.S.S.R. can most favorably compete." These space initiatives would fall predictably into four main categories: lunar and interplanetary probes; manned spaceflight activities; strategic photo-reconnaissance; and unmanned robotic missions for near-Earth scientific explorations. As in the past, the NIE observed, the Soviets would continue to rely on military boosters, experiment with new techniques for rendezvous and docking, and push ahead with plans for a "new large booster with the thrust of two million pounds." The latter project, a veiled reference to what became known as the N-1 booster, signaled the intention of the Soviets to make a manned circumlunar flight its prime mission, with the first attempt coming as early as 1967.

One key conclusion of the NIE was that the Soviets could not make a manned landing on the moon before 1969.[33] "In sum," the NIE concluded, "the Soviet space program appears to be in a state of transition. While we can estimate technically feasible extension of all current projects, we believe that the Soviets do not have in hand the necessary economic and technical resources for undertaking all such projects simultaneously."[34] If still awed by their recent space triumphs, the NIE expressed for the first time a measured skepticism of Soviet capabilities, identifying certain built-in restraints and weaknesses associated with Moscow's future space initiatives.

No one in the West, however, fully comprehended the reversals that would bedevil Moscow's space fortunes after 1966. The Soviet Union continued to maintain tight secrecy over its space program, which complicated the work of the NIE analysts. For Americans seeking an accurate sense of Soviet capabilities, it was indeed like peering through a glass darkly. Beginning with Gagarin's epochal Earth orbit in 1961, the Soviets had engineered a sequence of space "firsts," cultivating in their wake the widely held image of technological superiority over their American rivals. As the Americans set in motion the Apollo program, tested successfully its huge Saturn rocket, and made concrete plans for a lunar mission, the Soviets found themselves struggling with severe crises in management and technology. Their space program, as if following the life cycle of a dying star, had expanded brilliantly with the Sputnik launches, the lunar probes, and the benchmark manned flights but would collapse in the quest to reach the moon ahead of the Americans.

The first serious blow—one with immense consequences—was the untimely death of Sergei Korolev in January 1966. The still anonymous Chief Designer had been a powerful central figure, who had worked tirelessly to keep the Soviet Union at the cutting edge of the new space age. He towered over his contemporaries as a forceful and innovative leader. Korolev shared with Wernher von Braun a lifelong vision of human exploration of space, effectively molding military missile programs to serve his ends. By the end of 1965, however, the burdens on Korolev had grown dramatically. With the highly competitive American Apollo program taking wing, Korolev was hard pressed to keep the Soviet space program viable. The pace became intense even for Korolev.[35]

His control of events in the Soviet Union was complicated, if not compromised fatally, by the advent of new rivals such as Vladimir Chelomei and Valentin Glushko—men who were determined to make their mark on the Soviet space program. Chelomei emerged as a talented rocket designer, the head of the OKG-52 design bureau. His main arena of work had been cruise and ballistic missiles, at the time a vital military priority for the Soviet Union. Chelomei won the patronage of Khrushchev, which made him a powerful figure in the many rivalries that beset the Soviet space program. Glushko was a one-time ally of Korolev and a renowned rocket propulsion expert. Always in the background was the Soviet military establishment, which on occasion mounted keen opposition to the space program, seeing it as a drain on finite resources that could be more properly allocated for research and development of missiles.

Partly owing to the fierce bureaucratic infighting, Korolev's health began to decline precipitously. The first signs of an impending crisis arose in December 1965, when he had taken time from his frenetic schedule to undergo medical tests for a bleeding polyp in his intestines. Initially, both Korolev and his doctors considered the existence of the polyps as essentially benign—a condition that could be cured through minor surgery. Korolev entered a Moscow hospital on January 14, 1966—two days after his 59th birthday—for surgery. Boris Petrovsky, a prominent Moscow physician and high-ranking government bureaucrat, took personal charge, expressing confidence that the polyps could be removed without difficulty, even telling his patient that he would be able to continue work for another 20 years. Petrovsky approached the fateful day in a relaxed mood, even scheduling a second operation that same day.

Korolev's daughter, Natasha Koroleva, has left a succinct account of what happened: "The surgeon . . . started at 8 A.M., used a rectoscope to remove the polyps endoscopically. My father hemorrhaged on the operating table, bleeding so severely that it couldn't be stopped. Petrovsky cut the abdomen to stop the bleeding and found a cancerous tumor which had not been visible before." At this juncture, Petrovsky found himself facing a severe crisis, and he frantically tried to save his patient's life. Korolev's daughter then recorded her father's final moments: "He [Petrovsky] began to remove parts of the rectum to take out the tumor. This took a long time. My father had an anesthetic mask on for eight hours. They should have put some kind of tube in his lungs, but his jaws had been broken in prison so they couldn't use the tube. His heart was not in good condition, and Petrovsky knew this. He completed the operation, but my father never revived."[36]

The passing of Korolev—sudden and catastrophic—left the Soviet cosmonaut corps in despair. His body was taken to the Hall of Columns in the Kremlin, a belated recognition of the Chief Designer's long service to the Soviet state. Here his body was laid in an open casket on a high pedestal, covered with flowers and surrounded by red and black bunting on the white columns in the ornate hall. As countless mourners passed the casket, the room was filled with the music of Tchaikovsky and Beethoven. For the first time the name of Sergei Petrovich Korolev was acknowledged in an obituary appearing in Pravda with a photograph. Other public honors in the Soviet Union followed, including a special parade in Red Square. His remains were eventually cremated and then buried in the Kremlin wall, a gesture conferring high honor by the Soviet leaders Brezhnev and Kosygin. Even the *New York Times,* on January 16, 1966, took note of Korolev's death, publishing an obituary and informing its readers that he had been the designer of Sputniks and manned spacecraft.[37]

In the wake of Korolev's death, Vasiliy Mishin assumed leadership of the troubled Soviet space program. Mishin had displayed considerable skill and talent as one of Korolev's chief deputies, earning a reputation for competence as an engineer. However, Mishin lacked Korolev's charismatic style and wide experience in the upper echelons of the Soviet bureaucracy. In terms of personality, Mishin was a cautious and methodical administrator, a man who failed to show the bold leadership style that had defined the Korolev years. Mishin did not move forthrightly to reduce the number of competing programs. However, he did continue to push the development of the N-1, the Soyuz series (even after the death of cosmonaut Komarov), and the experimental Zond and Cosmos lunar probes. The unmanned docking of the Kosmos 186 and Kosmos 188 in 1967 represented an important milestone under Mishin's tenure in the immediate aftermath of Korolev's death. Mishin would lead the Soviet space program into the mid-1970s.[38]

Korolev's legacy continued to shape the Soviet space program in the late 1960s. The N-1 moon rocket, whose origins could be traced to 1958, represented the Soviet challenge to the Saturn V. While the Soviet military—most notably in the Strategic Rocket Forces—opposed such costly space programs, Korolev had prevailed. In a parallel program, the L3 spacecraft was designed for a future mission to the moon atop the still untested N-1 booster. In Khrushchev's final months in power, he had also approved a second space initiative under Vladimir Chelomei, of OKB-52, to send two cosmonauts around the moon in the LK-1 spacecraft. Chelomei now moved forward with the powerful UR-500, or Proton booster, which emerged as a rival design to the N-1. The perceived threat of Apollo had prompted Khrushchev to approve these ambitious space projects. When Leonid Brezhnev toppled Khrushchev and assumed power in October 1964, he continued to offer support for these long-range space goals for the Soviet Union.[39]

In the final year of his life, Korolev had used the specter of Apollo to garner support in the higher levels of Soviet political leadership, but he faced continued opposition from his old rival Valentin Glushko. The two had engaged in a bitter dispute over the engine design for the N-1. Korolev realized that the new rocket required powerful engines to lift a projected payload of more than 100 tons into orbit. Consequently, he argued for the development of engines using liquid oxygen and liquid hydrogen propellants; in his view this super-cooled cryogenic mix offered optimal lift for the huge N-1 rocket. His powerful rival, Glushko, remained a strong believer in storable propellants such as nitric acid, arguing that cryogenic fuels were inherently unstable. Going back to 1962, Korolev had gained ascendancy in this fierce debate when his approach received official sanction. Glushko then stepped aside from the N-1 project, severing his ties with Korolev. In time, Glushko migrated to Chelomei's shop, where he worked on the RD-253 engine for the Proton booster. A desperate Korolev then turned to Nikolai Kuznetsov, a designer of turbo-prop engines for aircraft, to design and build engines for the N-1. Kuznetsov did possess a reputation as a talented designer, but he lacked the experience and knowledge of Glushko. The infighting continued into 1965, when Korolev maneuvered to have Chelomei's LK-1 project cancelled. In its place, Korolev opted for the L1, a modified Soyuz spacecraft, to be catapulted into a circumlunar orbit on Chelomei's three-stage Proton rocket, which was the first Soviet launcher not based on a military missile prototype.[40]

As the Apollo space missions drew closer to the moon, the venerable Soyuz would be used for yet another space spectacular. In January 1969, Soyuz 4 and Soyuz 5 docked in orbit, and two cosmonauts transferred from one spacecraft to the other in a dramatic spacewalk. Stemming from the Soyuz design, another key undertaking by the Soviets in the late 1960s was the Zond space series. Between 1968 and 1970 five Zond deep space missions were launched. The unmanned Zond series represented a remarkable feat of navigation and hinted at the long-range goal of the Soviets to undertake a circumlunar mission, if not an actual lunar landing. Zond 5 set a new milestone by orbiting the moon and returning to Earth in September 1968, just months before the historic Apollo 8 mission. The unmanned spacecraft had taken photographs of Earth while above the moon's surface. In November, Zond 6 took a series of remarkable pictures of the far side of the moon. The Zond probes suggested that the Soviets were seriously preparing for a manned lunar mission.[41]

The critical component for any manned Soviet lunar mission was the N-1 booster. With its L3 module attached, the N-1 rocket was a huge affair: It was 344 feet long and weighed more than 2,700 tons. The N-1 was nearly the same height as the Saturn V, which was 364 feet tall. The design and appearance of the N-1, however, was quite different with its contours and trusses between stages. The first stage of the N-1 consisted of a complex array of 30 engines with a projected thrust of 9 million pounds, or 4,500 tons; a second stage with eight engines; and a third stage with four engines—all powered by a volatile mix of liquid oxygen and kerosene. Hurried into production, with the rocket engines designed by the inexperienced Kuznetsov, the N-1 was an accident ready to happen. The dramatic test of the N-1 took place at Baikonur on February 21, 1969. The launch was scheduled without any full-scale ground tests of the engines. At ignition, the complex first stage roared to life, flames and smoke enveloping the pad as the resulting shock waves vibrated across the complex. But after 80 seconds the first stage engines shut down, causing the N-1 to crash 60 miles from the launch site. The ill-fated super rocket exploded in the remote desert outside Baikonur. A second launch followed on July 3, and the N-1 became engulfed in flames just seconds after lift off. The explosion of the second N-1 ended any Soviet hopes of making a circumlunar mission and keeping pace with the Apollo program.[42] Two additional launches of the N-1—also catastrophic failures—took place in June 1971 and November 1972, bringing down the curtain on this ill-fated experiment in rocketry.[43]

The failure of the N-1 represented a dramatic reversal for the Soviets, but the reasons for the Soviet space program's inability to keep pace with the Americans rested in large measure on non-technical factors. In particular, the bitter competition among the major Soviet designers begat great confusion. Competing technologies worked at cross-purposes, which hindered the space program at a critical juncture when the Americans were ready to assume a decisive lead. Chronic disarray was evident in the post-Korolev era. The Soviet space effort had evolved out of a military missile project. It garnered support from the highest echelons of political leadership, but the military sector maintained a persistent and often debilitating opposition. In addition, the planned nature of the Soviet economy, with its top-down allocation of resources, did not consistently encourage innovation or, more important, a logical and sustainable space program with incremental stages. The shroud of secrecy—pervasive and illogical—cast its shadow over the entire program. The Soviet space program recruited many talented engineers, designers, and cosmonauts, but they worked in a context of managerial chaos and waste. The Soviet regime, self-styled as being at the cutting edge of progress, became the midwife of a system characterized by gross inefficiency. In the end, it was the politics of the Soviet Union that dealt the space program its most severe blows.

THE FIRST LEAP

The NASA press release for Apollo 8, dated December 15, 1968, described the forthcoming mission in a straightforward manner: "The United States has scheduled its first mission designed to orbit men around the moon for launch December 21 at 7:51 A.M. EST from the National Aeronautics and Space Administration's John F. Kennedy Space Center, Florida." The Apollo 8 spacecraft with a three-man crew would depart the familiar environs of near-Earth orbit and cross 237,000 miles into outer space—passing outside the gravitational pull of the home planet. Before this time the farthest humans had ventured into space had been a modest 850 miles. NASA did not characterize Apollo 8 as a bold or quantum leap in manned spaceflight. Instead the language was matter of fact, suggesting that the mission would be "open-ended" and designed to allow the crew "to operate at lunar distances." One statement in the press release, however, did hint at the inherent dangers of the lunar excursion: "The mission will be carried out on a step-by-step 'commit point' basis." This vague hint of caution meant that the Apollo 8 spacecraft might be ordered home abruptly or placed on some "alternate mission" if deemed necessary. The mission's projected "Earth landing" was scheduled to take place some 147 hours after launch on December 27, at 10:51 A.M. EST.[44]

The official language describing the Apollo 8 lunar mission in no way captured the drama of the moment—the remarkable fact that humans for the first time would visit, though not land on, Earth's orbiting moon. The decision to launch such a mission was not part of the pre-set NASA Apollo blueprint; in fact, NASA redefined the mission for Apollo 8 in a curious, almost accidental fashion. In the course of 1968, myriad technical problems in the manufacture of the LM, especially its ascent rocket, had caused a series of delays. The Apollo 8 mission had been scheduled to test the LM, but this critical evaluation now appeared impossible before the end of the year. George Low, a talented NASA engineer, then proposed that the Apollo 8 mission be radically reset for a bold journey to the moon. Moreover, Low suggested an even bolder objective—a series of orbits around the moon rather than a figure-eight loop of the moon. Such a maneuver, he argued, was appropriate; astronauts would have to perfect this skill for any future lunar landing. After some deliberation, NASA administrator James Webb approved the stunning Low idea.[45]

A manned crossing of the vast empty void between Earth and the moon would be an unprecedented exercise, one filled with extreme risk and manifold technical uncertainties. Webb understood the less than ideal odds for success, but he nevertheless approved the bold idea. Webb's measured enthusiasm was shared by other high-ranking NASA leaders and the astronaut corps. In retrospect, the Low plan offered a welcome stratagem for NASA to make a historic leap forward. The year 1968 had been a grim interlude for the United States at home and abroad: the assassinations of Martin Luther King and Robert Kennedy, the outbreak of race riots and civil unrest, and the continued debate over an unpopular war in Vietnam. As for NASA, it faced the haunting specter that the Russians might soon launch their Soyuz spacecraft on one or more space spectaculars, perhaps even a lunar flyby.[46]

The new initiative would follow up on the highly successful Apollo 7 mission in October, which had buoyed confidence in the space agency. Apollo 7 had signaled that NASA was ready to resume manned spaceflights after a 20-month interregnum following the Apollo 1 fire. Just before the Apollo 7 mission three unmanned flights using Saturn V and the smaller Saturn IB rocket had been launched. With this backdrop, Apollo 7 astronauts Wally Schirra, Donn Eisele, and Walter Cunningham aimed to orbit the Earth for 11 days, equal to the time it would later take for a round-trip to the moon. Their mission was to thoroughly check out the Apollo spacecraft's command and service modules' readiness to fly Americans to the moon. If they succeeded, nothing would stand in the way of such a flight. If they failed, the chances of achieving Kennedy's lunar landing goal on time would greatly diminish.[47]

As if those stakes were not high enough, Apollo 7, carried aloft on von Braun's smaller Saturn IB rocket rather than the Saturn V, was to be the first manned Apollo mission. The flight turned out to be a total success, remembered for achieving every engineering and technical goal set for the crew and the spacecraft. On the human side, though, Apollo 7 did have its difficult moments. Mission Commander Schirra, a veteran of Mercury and Gemini flights, contracted a severe head cold early in the flight. An impatient Schirra, often joined by his crew, took a combative posture toward NASA mission controllers. He cancelled a scheduled television broadcast from orbit, telling the flight controllers, "We've got a new vehicle up here, and I'm saying at this point television will be delayed, without further discussion. . . ." The Apollo 7 crew nonetheless eventually redeemed themselves with NASA's Public Affairs Office by sending several lighthearted TV broadcasts back from their capsule.[48]

The willingness of Schirra to depart from the normal protocol of obedience to the mission script became evident when he reacted negatively to one series of scheduled tests, saying they were "ill-conceived and hastily prepared by an idiot." Schirra had a keen desire to focus on the critical goals, primarily the need to check out the redesigned Apollo spacecraft. As he later pointed out, his motivation was quite personal: His "next-door neighbor, Gus [Grissom], one of our [Mercury] seven . . . [and] two other guys I thought the world of" were killed in the Apollo 1 fire. As a member of the Apollo 7 crew he aimed to prevent such a tragedy from happening again, even as the team paved the way for a future lunar mission. The moon was the next stop.[49]

For the Apollo 8 mission, NASA selected three capable and experienced astronauts: spacecraft commander Frank Borman, command module pilot James A. Lovell, Jr., and lunar module pilot William A. Anders. The hard-driving leader, Borman, had graduated from West Point. He was 40 years old, the oldest of the trio. Lovell, an Annapolis graduate, possessed an engaging personality and during the Gemini program had accumulated vast experience in orbit. Anders was the rookie, and he had spent much of the previous spring in training with the lunar landing research vehicle.[50] The Apollo crew, in many ways, reflected the prevailing personal profile of a NASA astronaut: a veteran test pilot, standing 5 feet 10 inches tall, weighing 160 pounds, with an average IQ of 141.[51]

The Apollo 8 mission was scheduled for December 21, a date selected to coincide with a new moon phase. The deliberate timing would allow the crew to orbit above the Sea of Tranquility during a lunar sunrise. This particular area was being considered as a future landing spot. The Apollo 8 crew would have the unique opportunity to study and photograph up-close the jagged and crater-filled lunar terrain.[52]

To negotiate the vast distance from Cape Kennedy to the moon, the remarkable Saturn V rocket had been assembled in the huge Vehicle Assembly Building (VAB), the largest structure in the Cape Kennedy complex. The Saturn was then transported 3.5 miles to its launch pad on a treaded crawler-transporter, moving down a gravel-covered, 110-foot-wide roadway. The unique space age transport, weighing six million pounds, moved at one mile per hour. At the time they were built, the crawlers were the largest tracked vehicles ever made, with a top deck roughly the size of a baseball infield.[53] The crawler also moved into place a mobile service structure used for the final servicing of the Saturn V. Each of the twin octagonal-shaped launch pads for the Saturn V contained a gigantic, heavily reinforced concrete pad and a flame deflector in the shape of an inverted V. The deflector channeled the river of exhaust flames from the Saturn V's first-stage engines into flame trenches on both sides, away from the pad. The trenches were lined with a ceramic surface that could withstand temperatures of nearly 2,000°F and flames moving at hypersonic velocity. A mini-tank farm along the pad's perimeter stored the Saturn's liquid oxygen and liquid hydrogen fuels. The Launch Control Center for Saturn V missions was housed in a multistory building adjacent to the VAB, located 3.5 miles from the launch pad for safety reasons.[54]

The five F-1 engines comprising the Saturn V's first stage provided a combined thrust of 7.5 million pounds—160 million horsepower. Each F-1 engine bell alone measured 12 feet in diameter. During the firing of the first stage, the Saturn V burned half a million gallons of kerosene and liquid oxygen in less than three minutes.[55] Given the enormous size of the Saturn rocket engines and the volatile mix of cryogenic fuels, NASA took care to plan for any "worst-case scenario." One palpable fear was an explosion—always a potential threat in liquid propellant rockets. Any explosion of a fully fueled Saturn V might generate a blast with the force of 500 tons of TNT. Such a conflagration would be enormous, about 1/26th the size of the atomic bomb that destroyed Hiroshima in 1945. A sudden explosion might create a fireball 3,000 feet wide, powered by the rocket's fuels and oxidizers: super-cooled liquid oxygen, thousands of gallons of kerosene, and extremely cold liquid hydrogen. For the astronauts atop the Saturn V, the only abort option was the launch escape system, designed to blast the command module away from the launch pad at high speed. To minimize collateral damage, the Saturn V launch pads had been built on isolated Merritt Island, with its generous swath of uninhabited swampland and beaches.[56]

Launch day came on Sunday, December 21, with liftoff at 7:51 A.M. from Pad 39-A. The Launch Control Center was staffed with a vast array of engineers, technicians, and mission control personnel gathered to oversee and monitor the launch of the Apollo 8 spacecraft. The mission created great excitement within NASA; everyone fully realized that this would be a truly historic mission. The long countdown proceeded in the scripted manner; there were no glitches. At T minus 15 minutes, with the final status checks under way, Borman and his crew realized that the liftoff for the moon was now imminent. The final countdown proceeded at a steady pace, with all eyes glued on the Saturn V. The moment of ignition came in a dramatic way, as the F-1 engines fired, spewing a torrent of fire and smoke at the foot of the rocket. The Saturn V lifted from its moorings as the hold-down clamps were released. The huge rocket quickly gained speed and altitude, going supersonic just after 40 seconds. A little more than 11 minutes into its flight, Apollo 8 was in orbit at 119 miles above Earth. The spacecraft was now traveling at 17,400 miles per hour.[57]

Late in the second orbit, NASA Mission Control in the voice of Michael Collins gave the go-ahead for translunar injection (TLI): The third stage rocket was fired, boosting the Apollo 8 spacecraft out of Earth orbit toward the moon. This firing propelled the spacecraft outward at 24,226 mph, in what NASA called "free-return trajectory," so that if no further maneuvers occurred, the Apollo 8 spacecraft would sweep around

Lift-off! Saturn V rocket carrying Apollo 11 leaves the Cape.

OPPOSITE: *Armstrong, Aldrin, and Michael Collins head for the launch pad—and the moon.*

PAGE I: *Buzz Aldrin salutes the Stars & Stripes on the lunar surface, Apollo 11, July 1969.*

PAGE 2-3: *Armstrong and Aldrin train for Apollo 11, Manned Spacecraft Center, Houston, April 1969.*

*En route portraits of Aldrin (*OPPOSITE TOP*), Armstrong (*OPPOSITE BOTTOM*),*
*and Collins (*ABOVE*) aboard Apollo 11, July 1969.*

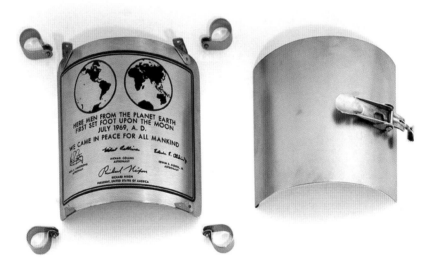

TOP: *For the ages: Aldrin's boot print on the lunar surface.*

BOTTOM: *Plaque left on the moon to mark humankind's first visit.*

OPPOSITE: *Aldrin descends from the lunar module.*

Aldrin's shadow on the lunar surface.

OPPOSITE: *Lunar module prepares to dock with command module as Earth "rises" in the background.*

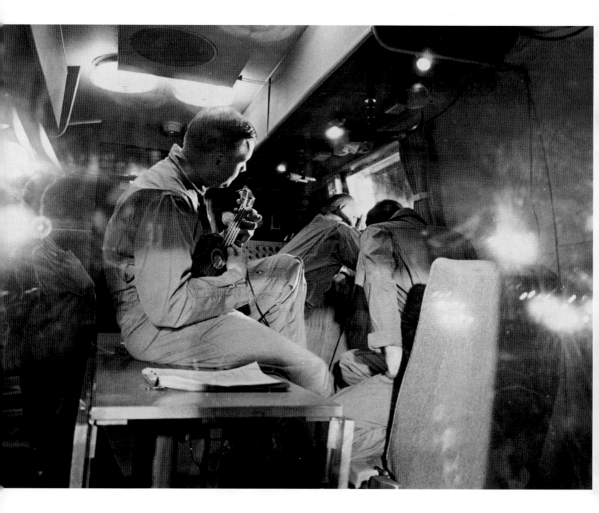

Apollo 11 crew in post-mission quarantine facility; Armstrong strums a ukulele.

"Did the photographs turn out?" Apollo 11 crew reviews trip film, August 1969.

Moment in the sun: New York City tickertape parade honors Apollo 11 astronauts, August 1969.

A triumphant von Braun borne on the shoulders of city officials in Huntsville, August 1969.

NEXT PAGE: *Apollo's iconic "earthrise" photographs offered a new sense of the planet's fragility and uniqueness.*

the moon and then make a direct return for home after 136 hours in space.[58] However, the script for the mission called for 10 orbits of the moon for navigation and photography work, meaning the crew would return to Earth 147 hours after launch.

Crossing the forbidding gulf that separated Earth from the moon turned out to be a peculiar experience for the crew. They could not sense any movement, notwithstanding their remarkable acceleration from Earth orbit. The crew depended on Mission Control for clear and accurate data on their position in space. They experienced the unique sensation of seeing Earth recede, becoming smaller and smaller to the eye. They were aloft in a world of weightlessness, with no up or down, no day or night, almost a sense of being stationary. For long hours the moon remained out of sight; the object of their journey would appear only at the end. The Apollo 8 crew speeding across the translunar void was forced to realize that all of humankind—except for those in their capsule—were to be found on that small blue planet shrinking in the distance. The passage out of the gravitational pull of Earth, roughly three days outward bound, marked an important benchmark on the lunar flight. Soon they passed an invisible line into the moon's gravitational sphere. Those who made that crossing have reported a keen sense of isolation, dislocation, even vulnerability.[59]

One of the more daunting tasks facing Project Apollo was navigation: traveling across 237,000 miles to the moon and back. The equations for plotting Apollo 8's course were far more complex than calibrating the relationship of two objects in space. In this case three objects—Earth, the moon, and Apollo 8—were all moving, and each body, varying in size, was influenced by the gravity of the other. The relative positions of each to the other had to be calculated with great precision. Only the advent of computers offered a solution to the "problem of three bodies," and even then only on a highly specific lunar mission-to-mission basis. The ability of NASA Mission Control to predict where the moon, Earth, and the Apollo 8 spacecraft would be relative to each other at all times depended on high-speed computers. Without them to make the many calculations about direction and velocity, the Apollo mission would not have been possible.[60]

Because the Apollo program unfolded during the Cold War, Apollo 8 faced an unknown factor. How would the Soviets react to an unparalleled American success? As astronaut Michael Collins observed, ". . . the Russian program was hidden from view, secret, and mysterious, and if our side knew what was going on, the information never trickled out of the CIA files down to us working troops in Houston."[61] On a personal level, during a rare encounter between astronauts and cosmonauts, Collins had met Pavel Belyayev and Konstantin Feoktistov at the Paris Air Show in 1967 and later recorded that his Soviet counterparts were "good fellows, indeed."[62] Yet some NASA alarmists feared that the Soviet Union might jam navigational information sent from ground-based computers to Apollo spacecraft, a move that might compromise a journey to the moon. An innovation to circumvent this problem was to put computers on board the Apollo command module and the lunar excursion module. As the program unfolded, though, computers on the spacecraft proved to have a much more realistic, even essential, mission role. In the end, the on-board system was so thoroughly integrated with the spacecraft that it was considered "the fourth crew member." Beyond navigation, its functions included management of several guidance components and other systems. As for the LEM's computer, it provided autonomous landing, ascent, and rendezvous guidance.[63]

The command module's computer was programmed with coded instructions for the flight to the moon on erasure-proof, plastic-encased bands of magnetized wire. Though primitive, it was capable of getting the astronauts to the moon and back.[64] Despite such significant roles, Apollo's on-board computers were nonetheless extremely limited compared to the capabilities of even today's desktop models. For example, the Apollo guidance computer weighed 70 pounds and was housed in a three-by-five-foot flat box. It had a one-megahertz processor, one kilobyte of random-access memory and 12 kilobytes of read-only memory. Today's typical desktop computer offers a thousand times the processor speed and perhaps 500,000 times the RAM. Hard drives have replaced ROM, providing millions of times the capacity.[65]

The Apollo 8 mission was too complex for the crewmembers to master every aspect of the journey. Accordingly, each astronaut specialized in certain tasks. For example, in addition to his mission commander's role, Borman had responsibility for steering the command module through reentry if the computer failed. Anders thoroughly mastered the command module's systems. Lovell handled the complex navigation tasks, for example employing a sextant to verify that the spacecraft was, indeed, on course.[66]

After a three-day journey, an intensely dramatic moment arrived. The astronauts made their first visual contact with the moon—now a huge, looming celestial body in close proximity. As true pioneers, the crew now executed the lunar orbit insertion (LOI) maneuver, where the combined CSM's service propulsion system (SPS) engine fired to slow the Apollo 8 and allow it to enter an orbit around the moon. The SPS engine had only one thrust chamber and one exhaust nozzle—on this thin reed the safety and success of the mission resided. The LOI was also undertaken without radio contact with Mission Control.[67] As the crew went through their checklist for the LOI burn, they suddenly found themselves enveloped in darkness as the spacecraft swung around and fell into the shadow of the moon. Anders looked out on the blackness and, for the first time, saw the sky filled with countless stars. Andrew Chaikin in his celebrated book, *A Man on the Moon,* vividly captures this extraordinary scene: "He craned toward the flat glass to look back over his shoulder . . . and he noticed a distinct arc beyond which there were no stars at all, only blackness. All at once he was hit with the eerie realization that this hole in the stars was the moon. The hair on the back of his neck stood up."[68]

The Apollo 8 spacecraft now entered an orbit 69 miles above the lunar surface. The whole burn had created great anxiety back home. Yet the SPS engine—to the relief of Mission Control—fired on schedule and without mishap. The fears that Apollo 8 might fall into a dangerous orbit or even crash into the moon gave way to jubilation. Because the successful lunar orbit insertion was executed on Christmas Eve, the coincidence led to one of the most memorable episodes in the Apollo program—one that would be shared with countless millions on Earth. A television broadcast had already been scheduled for December 24. The crew prepared for the broadcast while taking in a moving sight—Earth rising beyond the moon. The telecast began with Borman describing the events of the day, the crew at work on its many tasks. He then described the desolation of the moon. Lovell joined in and spoke of orbiting a "grand oasis." Anders chronicled the passage of Apollo 8 above lunar landmarks such as the Sea of Crises, the Sea of Fertility, the Marsh of Sleep, and finally the Sea of Tranquility. Then the crew read passages from the creation story in the Book of Genesis. Borman then concluded: "And from the crew of Apollo 8, we close with, good night, good luck, a Merry Christmas, and God bless all of you, all of you on the good Earth."[69]

The Christmas Eve telecast would linger as an enduring episode of Apollo 8, and the photography generated on the mission became a lasting legacy of the first human flight to the moon. The photos taken by the crew framed a new perspective not just on the moon, the first object of the camera, but on Earth itself. The key visual breakthrough came with the unplanned "earthrise" imagery. This stunning perspective of Earth came on Apollo 8's fourth lunar orbit, as it passed behind the far side of the moon. As Anders looked out the window, he suddenly exclaimed: "Oh, my God. Look at that picture over there." Borman then asked, "What is it?" Anders' simple rejoinder expressed the emotion of that unprecedented moment: "The Earth coming up. Wow, is that pretty." Out the window, the first "earthrise" was observed by humans as they orbited above the bleached gray lunarscape. The contrast between the stark and barren uniformity of the moon with Earth was clear and dramatic to the astronauts. The blue-and-white globe emerging above the lunar horizon possessed vibrant and deep colors, set against the backdrop of the black infinity of space. The iconic photos taken by Anders were acclaimed as among the most important in history, giving rise to a new human appreciation of the unique fragility of their planet. Anders later recalled his feelings as he viewed his home from the remoteness of space: How tiny Earth was in the cosmos. Humans had come a great distance to explore the moon, he realized, and what we really discovered was a new viewpoint of Earth.[70]

The return home for Apollo 8 began with the tension-filled trans-Earth injection (TEI). The critical maneuver rested on one engine and one firing; the failure of the engine would mean being stranded in lunar orbit with a finite supply of oxygen and life-support capabilities—and certain death for the crew. To the great relief of Mission Control and countless observers on Earth, the TEI proved effective, propelling the Apollo 8 crew homeward. The journey back to Earth took three long days, a reminder of the enormous distances covered by the pioneering Apollo 8 spacecraft. Reentering Earth's atmosphere at high speed, Apollo 8 followed a precise angle. Once through the heat and pressure of reentry, the parachute system deployed properly and on time, allowing for a safe splashdown. Apollo 8 marked one of the most stunning and precedent-making space missions of the era.[71]

With the success of Apollo 8, full-scale testing of the complex lunar module (LM) moved forward as the next high-priority goal. Perfecting the LM became the final hurdle before any lunar landing could be attempted. Given the complexity of the LM design, two Apollo missions were set aside for exhaustive testing. On any future lunar mission, the LM would be deployed from its third-stage adapter section and then docked with the command module. From here a crew would enter the LM and then make a powered descent to the moon surface. Apollo 9 offered the first occasion to test the LM systems—in Earth orbit, where advanced rendezvous techniques could be developed and refined.[72] Apollo 10 then followed with a full dress rehearsal, taking the LM and its two astronauts down to only 50,000 feet above the moon. Assuming success in both tests, a July 1969 lunar landing was then planned. The staggering complexity of both missions made success far from certain.

Apollo 9 was launched on March 3, 1969, on an Earth-orbital mission of 10 days. Jim McDivitt served as mission commander. He came to his assignment with considerable experience, having been an Air Force pilot in Korea and a member of the Gemini 4 crew. McDivitt was joined by David Scott, a former Air Force test pilot and participant on Gemini 8, and Russell "Rusty" Schweickart, also a former Air Force fighter pilot and a former MIT research scientist.[73] Once in orbit 119 miles above Earth, the Apollo 9 crew initiated a complex ballet to deploy the LM and then dock it with the command module. During the launch, the LM had been stored in an adapter section of the Saturn V's third stage. The first step called for the command and service modules (CSM), when locked together, to separate from the third stage of the Saturn V launch vehicle and then drift away a few yards to execute a 180-degree turn. The next step was opening the adapter section's four conical garage-door-like petals, which separated and then drifted off into space. Inside was the LM, ready for extraction and docking. Scott positioned the CSM directly nose-to-nose with the LM, and brought the two vehicles together, with the CSM's nose-mounted docking probe fitting neatly into the docking port on the LM's roof. Automatic docking latches sealed the ships together. Next came the extraction of the LM from its garage. Once the tunnel joint between the two modules became rigid and secure, the entire Apollo spacecraft was spring-ejected from the Saturn third stage. Scott then used the CSM's reverse thrusters to back away.[74]

During the 10-day mission, the crew succeeded in testing the CSM, which they named *Gumdrop,* and the LM they called *Spider.* It was the kind of mission test pilots dreamed about: with Scott in the CSM and McDivitt and Schweickart aboard the LM, they undocked the spacecraft. The two astronauts flew the *Spider* for six hours, venturing more than 100 miles from the command module before firing their engine to head back and rejoin Scott. Redocking required precise use of the orbital rendezvous techniques laboriously perfected on the Gemini missions. This was essential practice for the day when the command module and the lunar module would have to find each other and rejoin above the surface of the moon.[75]

Apollo 10 was next up in the LM test program. On May 18, 1969, the mission began an eight-day journey to the moon. Once in lunar orbit, Apollo 10 undertook one of the most dramatic and dangerous experiments in deep space—the powered descent of the LM with two astronauts aboard to within 50,000 feet of the lunar surface, using a fully functional lunar module. The crew consisted of mission commander Tom Stafford with Gene Cernan in the LM and John Young in the command module. The crew was experienced and talented, having no less than five Gemini missions among them. The selection of three space veterans, no doubt, reflected the crew's own belief that the trip would be more demanding than all the previous Apollo missions combined.

Like Apollo 8 before them, Stafford and his crew entered translunar injection, but this time with a major difference. They were taking an LM they called *Snoopy* with them to the moon, mated with *Charlie Brown,* their command module.[76]

On May 22, after boarding the lunar lander and detaching it from *Charlie Brown,* Stafford and Cernan fired the LM's descent rocket at an altitude of 69 miles above the moon. Their descent orbit carried them down to around the prescribed 50,000 feet. At that altitude they saw huge boulders scattered across the landscape below, with some appearing to be the size of five-story buildings. Stafford later learned they were 10 times larger. "Houston, this is *Snoopy!* We is Go, and we is down among, 'em, Charlie," an excited Cernan radioed Mission Control as he and Stafford watched their first earthrise.[77] Flying through blackness nine miles over the surface, the LM crew also surveyed the Sea of Tranquility, planned as the first lunar landing site. They reported that while some safe areas existed, much of the site was covered with boulders and craters.

The *Snoopy* crew had a brief but terrifying scare on their way back to link up with the command module. The LM's automatic guidance system caused the maneuvering thrusters to fire wildly. The LM began tumbling out of control, causing Cernan to yell, "Son of a bitch. . . . What the hell happened?" The crew regained manual control of the spacecraft after eight long seconds; the problem turned out to be easily correctable.[78]

The remainder of the voyage went as planned. America now had sent six astronauts to the moon and safely returned them. Nothing stood in the way of a lunar landing. Yet, there was the lingering fear that the Russians might upstage the United States at this critical juncture. While their intentions were shrouded in mystery, the launch of Zond 5, an unmanned photographic mission from lunar orbit, prompted anxiety in the American space program. Sending a highly automated probe into lunar orbit and returning it successfully, in the view of Michael Collins, "represented a hell of a capability and made a lot of NASA people edgy."[79] The eleventh-hour launch of another unmanned Soviet mission, Luna 15, at the time of the actual lunar landing in 1969 would prompt similar forebodings.

A LANDING ON THE SEA OF TRANQUILITY

The Apollo 11 launch took place on July 16, 1969. The moment possessed a cosmic aura of its own. NASA simply stated in a press release on the eve of launch date that the Apollo mission would "perform a manned lunar landing and return." In a narrow sense, the extraordinary space venture would achieve the "national goal" of a successful manned lunar landing by the end of the decade. Yet it would accomplish a much greater dream, one NASA expressed in an understated and cautious manner: "If the mission—called Apollo 11—is successful, man will accomplish his long-term dream of walking on another celestial body."[80]

The audacious undertaking called for a three-man crew to make the long journey to the moon, and two astronauts to land and explore the moon's surface. They would collect lunar samples, deploy scientific experiments to transmit back to Earth important data, and photograph the landing site and the surrounding environs. To mark the historic occasion, the Apollo 11 explorers would leave a plaque bearing a map of Earth with the following inscription:

HERE MEN FROM THE PLANET EARTH
FIRST SET FOOT UPON THE MOON
JULY 1969 A.D.
WE CAME IN PEACE FOR ALL MANKIND.

The crew selected for the Apollo 11 mission brought considerable experience and technical expertise to their assignment. The commander of Apollo 11, Neil A. Armstrong, a veteran test pilot, received the coveted nod to take the first steps on the moon's surface. He would be joined by lunar module pilot Edwin "Buzz" E. Aldrin, Jr., for the high-risk lunar landing. Michael Collins would serve as command module pilot. All three astronauts were in their late thirties at the time. A former Navy pilot with combat experience in the Korean War, Armstrong had come to NASA as a civilian; his crewmates were both experienced Air Force officers. Collins, an articulate veteran of the Gemini program, has left insightful descriptions of his two fellow Apollo crewmen. Armstrong, as described in Collins's own memoir, was cool under pressure, hardworking, and seemingly nonchalant— "very much on top of a complex and rapidly changing situation." In Buzz Aldrin, the second man to walk on the moon, Collins saw another purposeful colleague, who was "all business" during the training for their historic mission. However, Collins remembered that Aldrin—"if generally quiet and incapable of small talk"— would expound endlessly on his "pet projects." Both men, in Collins' view, were talented and creative engineers.[81] Aldrin held a Ph.D. in astronautics from the Massachusetts Institute of Technology.

The launch of Apollo 11 was for NASA a tense event that passed without any technical glitch—again demonstrating the impressive power and reliability of the giant Saturn V rocket. The Apollo 11 crew had named their command module *Columbia* and their lunar landing vehicle, the *Eagle*. Once in orbit, the crew of Apollo 11 received the welcome order for the translunar injection, and the third stage of the Saturn V engine reignited to boost the spacecraft outward on a trajectory to the moon. During its three-day journey Apollo 11, as with other lunar journeys, repeated the thermal-control maneuver of past lunar missions. In it the spacecraft was positioned broadside to the sun and then rotated slowly, in the words of one astronaut "like a chicken on a motorized barbecue pit."[82] The passage to the moon followed the script of two earlier missions, Apollo 8 and Apollo 10. Upon reaching the moon, the Apollo 11 spacecraft slowed in a two-burn firing to allow its capture by the moon's gravitational field. Once in its new orbit, Apollo 11 followed a nearly circular path around the moon, an orbit averaging 60 miles above the lunar surface. The transposition and docking of the LM followed, repeating the well-rehearsed maneuver of extracting it from the Saturn's third stage for docking with the command module. On the fourth day, all was ready for a descent to the moon's surface.[83]

The encounter with the moon prompted a sense of awe in the Apollo 11 crew. "To begin with," Michael Collins reported in his memoir, "it is huge, completely filling our window. Second, it is three-dimensional: The belly of it bulges out toward us in such a pronounced fashion that I almost feel I can reach out and touch it. . . . It is between us and the sun, creating the most splendid lighting conditions imaginable." Once observed up close, the ever-shifting perspective of sunlight gave the moon a special character: "The sun casts a halo around it," Collins reported, "shining on its rear surface, and the sunlight which comes cascading around its rim serves mainly to make the moon itself seem mysterious and subtle by comparison, emphasizing the size and texture of its dimly lit and pockmarked surface. . . . This cool, magnificent sphere hangs there ominously, a formidable presence without sound or motion, issuing us no invitation to invade its domain."[84]

On Sunday, July 20, the Apollo 11 crew prepared for a descent to this forbidding domain. The procedure began with the release of the LM, quickly followed by a two-burn sequence to power Armstrong and Aldrin downward toward the lunar surface. Though practiced repeatedly in Earthbound simulators, the actual moon-landing maneuvers entailed many risks for the crew. The *Eagle* had to land in a precise spot, selected in advance, to assure safety. Initially, Armstrong and Aldrin assumed the odd position of heads up and feet forward; this awkward stance was maintained during the powered descent to the lunar surface. At this juncture they peered out their windows into a black void. Then, on signal, the LM rolled over and Armstrong began a tension-filled quest for the proper landing site. The two astronauts encountered what appeared to be a "program alarm," but Mission Control responded with instructions to ignore the warning.

The LM now began its final descent to the Sea of Tranquility. Armstrong peered out the window of the LM to oversee the climactic computer-run descent program. Nearby in the close quarters of the fragile LM, Aldrin monitored their progress by yelling out periodic readings on altitude and velocity. Soon the targeted landing zone came into view—clearly visible to Armstrong at approximately 1,000 feet altitude. For Armstrong, the planned landing area now looked problematic. Suddenly, he spotted a large crater, surrounded by boulders and rocks, some quite large and imposing. He then pitched the LM forward until it was nearly level, gliding over this field of boulders at 350 feet. Armstrong had acted swiftly, even as Aldrin continued his audible readings: "Six hundred feet, down nineteen [feet per second]. . . . Four hundred feet, down at nine. . . . Two hundred feet, 4 1/2 down. . . . One hundred feet, 3 1/2 down, nine forward. Five percent [fuel remaining]. . . . Forty feet, down 2 1/2, kicking up some dust. . . ." Houston chimed in: "Thirty seconds. That's how much fuel they have left. Better get it on the ground, Neil." Within an instant that seemed like an eternity, Armstrong deftly steered the *Eagle* to a soft landing—although the braking thrust of the LM generated a huge cloud of dust. "Contact light," Aldrin reported exuberantly, signaling confirmation of a safe touchdown. At Mission Control the tension had reached a fever pitch. "We copy you down, *Eagle*," Houston announced to the Apollo 11 crew—a message, in Collins' words, that was "half question and half answer." Finally, Armstrong confirmed the space triumph—"Houston, Tranquility Base here. The *Eagle* has landed."[85]

Later that night Neil Armstrong descended from the hatch, moving deliberately down a small ladder, and stepped off onto the lunar surface. The moment was electric and unforgettable—the first excursion by a human being on the moon. The event was televised to a vast global audience on Earth. Armstrong's brief words captured the epic nature of the moment, "That's one small step for man, one giant leap for mankind."

Aldrin joined Armstrong on the lunar surface. Above them was the canopy of a pitch-black sky, broken only by the intense light of the sun. Around them was, as Buzz Aldrin aptly described it, the "magnificent desolation" of the lunar surface—jagged, chalky gray, and forbidding. Lightly tethered by the moon's one-sixth gravity, the two astronauts planted an American flag, set up seismic sensors and other scientific packages, and gathered rock samples. Armstrong even indulged his curiosity about an 80-foot crater he and Aldrin had passed over just before landing. He ran over to inspect it for a brief moment and to shoot some spectacular photographs, with the lunar module visible in the background.

The astronauts' nearly two-and-a-half-hour EVA proved to be as exhausting as it was exhilarating. There were still many preparations, and scheduled rest, before their assent. Their stay on the moon's surface proved to be short—22.5 hours in all—but it constituted an extraordinary and unprecedented episode in the saga of human exploration.

Once reunited in the cocoon of the Apollo 11 command module, the three astronauts shifted their attention to the task of navigating home. In the distance—across an enormous void—was a beckoning blue planet, alone and unique. Now began their triumphal voyage home.

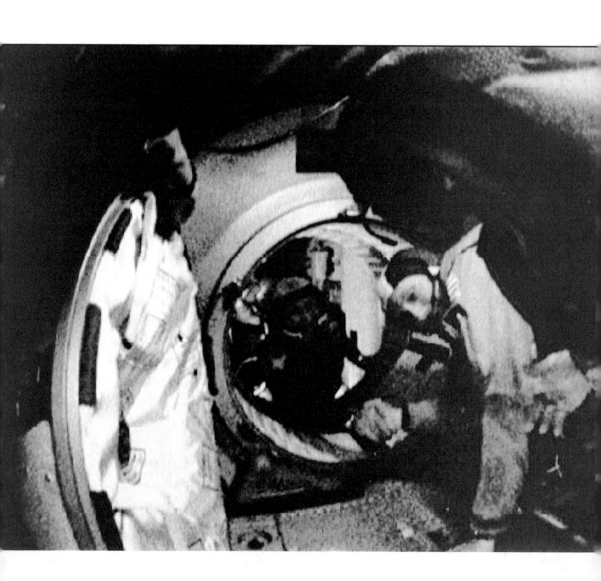

EPILOGUE

Old Visions, New Realities

The Apollo 11 mission stands out as a defining moment in the history of the 20th century—if not of the millennium that soon closed. Earthlings boldly stepping onto the surface of the moon represented an extraordinary moment in the human experience. Televised coverage of Neil Armstrong and Buzz Aldrin exploring the distant lunar surface evoked a sense of awe, even magic. Most earthbound observers felt that the Apollo 11 mission represented more than a raw display of America's formidable technological and engineering prowess. Some sought to cast the entire saga of space travel in universal terms as a triumph for civilization—a welcome counterpoint to a century filled with wars, revolutions, and vast human suffering. Still others chose to interpret it more narrowly as a welcome victory for the United States in the Cold War. In the summer of 1969, Americans found themselves bedeviled by terrestrial concerns: a divisive war in Vietnam, racial conflict, and growing social discord. The 1960s had signaled a time of profound cultural change

OPPOSITE: *Astronaut Stafford and cosmonaut Leonov meet after link-up of Soviet and American spacecraft, July 1975.*

in American life. For many, this immediate context of change and uncertainty dampened any urge to celebrate.

Ironically, the Apollo 11 mission—and the five successful Apollo lunar treks that followed it—did not herald a new age of manned space exploration for the United States. The fabled mission to the moon had ended an era, not opened one. As public interest in space waned and other competing national priorities arose, budget allocations for NASA were sharply reduced. Facing this wave of fiscal austerity, the American space program took on a variety of new incarnations, with manned missions restricted to near-Earth orbits. Some of the new orientation reflected genuine, if neglected, NASA programs. Robotic missions to explore the solar system soon received renewed emphasis. For ardent space visionaries, the ancient dream of a human visitation to Mars would persist, but the grand enterprise had been deferred indefinitely.

The Soviet Union faced its own limitations. Still a superpower, the communist state nonetheless had to contend with powerful internal stresses and contradictions. This hidden dynamic would bring about the demise of the Soviet regime within a generation, even though such a fateful turn of events seemed implausible in 1969. The Soviet Union had keynoted the space age with the launch of Sputnik, Yuri Gagarin's fabled orbital flight, and a sequence of other dramatic "firsts" in space exploration. The highly secretive spaceport at Baikonur had served as a platform to fire an array of advanced rockets and pioneering robotic space probes. Yet these formative space spectaculars had not assured any permanent lead for the Soviets in the space race. In the quest to launch a manned mission to the moon, the Soviet Union proved itself to be lacking in the organizational and technological means to compete. In retrospect, the Soviet Union's debut as a space-faring nation has to be regarded as impressive in its own right. Not only that, the Russians established a presence in space that would endure beyond the life span of the Soviet regime.

Moscow's propaganda organs had greeted the space age with a flurry of hyperbole, proclaiming that the Soviet Union's early triumphs reflected the powerful and relentless dynamic unique to communism. In reality, both the Soviet Union and the United States had embraced technology as an effective tool to sustain a competitive edge in the Cold War. The Soviet regime benefited from its talented group of designers, engineers, and scientists who established an early momentum. This edge, however, did not long endure once the American space program acquired a clear mandate, proper footing, and lavish funding. By contrast, the weaknesses of the Soviet space program became evident, a lack of forward progress made more severe by a cumbersome bureaucracy belabored by internal strife, waste, and inefficiency. By the late 1960s, the Soviets had become essentially spectators in the high-profile lunar quest. The failure to keep pace with Americans, however, did not spell the end to Russia as a space-faring nation: In the post-Apollo era Russian rocketry and orbiting space stations would sustain—even define—the new space age.

Looking back, this new epoch of exploration had been characterized by fast-paced technological breakthroughs. The Cold War with its military missile programs had been a catalyst that fueled this extraordinary innovation on many fronts. When the Soviets placed the first artificial satellite into orbit in 1957, they ignited a phenomenal sequence of technological advances on both sides. Rocketry and spacecraft design advanced quickly, allowing for manned orbital flights beginning with Yuri Gagarin in 1961. For the Apollo missions in the late 1960s, the United States routinely launched its huge and reliable Saturn V rocket—truly one of the technological wonders of the space age. The era became a showcase of innovative spacecraft designed for long-duration flights, space rendezvous and docking, and landing on the moon's surface. The quickening pace of space exploration was also measured in the early robotic space probes, beginning with Luna 1, launched in 1959 by the Soviet Union. It became the first man-made object to leave the gravitational field of Earth, fly past the moon, and enter an orbit around the sun. Subsequent deep-space probes by both nations revealed remarkable data on the atmosphere of Venus and the surface of Mars. The era required visionaries and talented engineers on both sides. The organizational demands were daunting and calendar-driven—and here NASA prevailed by demonstrating a prodigious capacity to meet the stated national goal of a manned lunar mission before the close of the 1960s.

In the immediate aftermath of Apollo 11, the United States launched six Apollo lunar missions. Apollo 12 followed in November 1969 with Alan Bean and Pete Conrad making an impressive lunar landing in the Ocean of Storms—a mere 600 feet from the Surveyor 3 unmanned probe that had landed on the moon two years before. A near tragedy with Apollo 13 followed in April 1970: The explosion of an oxygen tank compelled astronauts Jim Lovell, Fred Haise, and John Swigert to circle the moon without a landing, using their lunar module as an improvised life-support system to reach home. Nearly a year followed before Apollo 14 took Alan Shepard and Ed Mitchell to the Fra Mauro lunar highlands. In the Apollo 15 mission, July 26–August 7, 1971, David Scott and Jim Irwin explored the lunar surface for three days, making use of a lunar rover to range out from the landing site. One dramatic measure of success was the total of 18 hours of EVA that these two Apollo 15 astronauts accumulated on the moon. In April 1972, Charlie Duke and John Young on Apollo 16 reached the central highlands, exploring for three days. The final manned lunar mission, Apollo 17, came in December 1972 with Gene Cernan and Harrison Schmitt making use of a lunar rover to complete a long-duration exploration of the lunar surface.

NASA had originally planned 20 missions for the Apollo program, even a projected landing on the far side of the moon. In August 1969, just two weeks after the Apollo 11 landing, Wernher von Braun laid out an ambitious program for future space exploration to the Space Task Group, arguing for orbiting space stations, a permanent base on the moon, and ultimately a manned mission to Mars. However, the prospects dimmed for any bold program of manned flights in the near term. With the completion of the Apollo 17 mission the curtain descended, the consequence of the Nixon administration's decision to both reduce NASA's budget and to increase funding for the proposed Space Transportation System (STS), or space shuttle. These changes mirrored the shift in public support for NASA in the 1970s. Von Braun would leave NASA in 1972, and he died of cancer five years later.

Looking up at the moon 50 years after Sputnik 1, any observer is awe-struck by the extraordinary fact that 12 humans actually walked on its surface. In historical terms, Project Apollo represented an engineering triumph of the highest order. It ultimately achieved its lunar landing time schedule and delivered two successful roundtrips to the moon each year in 1969, 1971, and 1972. What's more, these feats were accomplished without today's highly sophisticated computers and advanced guidance systems. Since that time space visionaries have lamented the abandonment of lunar missions with all their potential for further human exploration of the solar system. For these space advocates, the space shuttle is a pale and uninspiring alternative to the heroic past. With the collapse of the Cold War, few have forecast, at least for the immediate future, another space race that would spark a new era of space exploration. International cooperation is the norm—with all its benefits and drawbacks.

One key shift in the post-Apollo 1970s was the desire by both the United States and the Soviet Union to build avenues for international cooperation. As early as 1972, the rivals in the Cold War signed an agreement for a joint space venture. In July 1975, this pact became a reality with the celebrated Apollo-Soyuz test project. Two spacecraft were catapulted into orbit, one from Cape Kennedy and the other from Baikonur. The Apollo and Soyuz spacecraft docked in space, marked by the handshake between Apollo commander Thomas P. Stafford and Soyuz 19 commander Aleksei A. Leonov. To achieve this remarkable thaw in the Cold War, both nations had collaborated in engineering a docking module to allow the historic rendezvous in outer space. This was the last mission to use the Apollo spacecraft. The American crew included Deke Slayton, a member of the original Mercury 7. The Soviets made their own polite bow to the past by sending up Leonov, the first person to walk in space and a veteran of the Vostok series. In a symbolic way, Apollo-Soyuz offered a fitting end to the space race. An epic rivalry had culminated in a joint spaceflight.

Before the Apollo-Soyuz mission, the United States signaled its dramatic shift to near-Earth orbital flights with the launching of Skylab, an experimental space station, in 1973. Fired into orbit atop a Saturn V rocket, a leftover from the Apollo program, the Skylab became a highly successful orbiting workstation. The concept of Skylab bore the imprint of von Braun and others in NASA committed to a permanent human presence in space. A crew of three astronauts worked in Skylab during a course of three missions. Fashioned from the third stage of Saturn V, the orbiting Skylab provided a generous 13,000 cubic feet of habitable space. Crews worked on a variety of experiments, using sophisticated instrumentation to gather a vast amount of data—a platform for both astronomical observation and the study of Earth's weather systems. An exercise in long-duration spaceflight, Skylab was nevertheless a temporary measure, no more than an experiment presaging a future permanent space station. It was abandoned in 1974 and then met a fiery end in the Earth's upper atmosphere in July 1979 after its orbit decayed. Fragments of Skylab fell harmlessly into Australia and the Indian Ocean.

At the same time, the United States was pursuing the development of the space shuttle, a reusable spacecraft. The launch of the first shuttle, *Columbia*, in April 1981 inaugurated this program. The spacecraft had been designed as a cost-effective alternative to the expendable launch vehicles of the past. As the program evolved, the space shuttles became the familiar vehicles of satellite delivery and retrieval, scientific research, and training of astronauts—now broadly recruited for their scientific and engineering specialties rather than narrowly for experience as test pilots. One near-fatal setback for the space-shuttle program came in 1986 with the breakup and fiery descent of the *Challenger* just 73 seconds after liftoff, with the loss of the entire crew. In February 2003, a second shuttle, *Columbia,* was also lost, killing the seven astronauts on board. But the space shuttle did achieve major triumphs as well. In 1990, the space shuttle *Discovery* placed into space the Hubble Space Telescope. Space shuttle astronauts then repaired and serviced the huge telescope riding above Earth's atmosphere and offering fascinating details on the structure of the universe. Despite its remarkable longevity, the space shuttle has prompted many critics, who argue that the reusable spacecraft did not deliver on its promises of quick turnarounds between missions and reducing the cost of delivery of materials to orbit. Some NASA critics have asserted the United States erred in relying on the space shuttle to launch many national security, military, and civilian satellites. This was done in the absence of any national space policy review, mixing military and civilian space operations on a single spacecraft.[1]

The Russian space program faced its own challenges in the altered context of the 1970s and 1980s. The venerable Soyuz spacecraft continued its remarkable lifespan as one of the most reliable manned space vehicles. First launched in 1967, the Soyuz evolved over time into the more refined Soyuz T and Soyuz TM types, becoming the baseline technology for Russian orbiting missions. More than 100 cosmonauts flew on Soyuz spacecraft during its long service life. In the 1970s, the Soviets launched Salyut, their first space station. Six more Salyut stations would follow, plus the durable, if chronically troubled, Mir, which remained in orbit between 1971 and 2001. Originally designed for a five-year service life, Mir endured for an additional decade. In some ways, Mir embodied the latter-day style of the Russian space program with a stress on cost cutting, austerity, minimal repairs, and the continued use of proven technologies.

Baikonur persisted as a viable space center in spite of periods of neglect and depressed funding. Since 1955, the launch center had been a major center for the testing of intercontinental missiles. It then expanded in the golden years as the Soviet Union's key center for manned and robotic space missions. In 1988, the Soviets launched the huge Energiya rocket with a Soviet space shuttle called Buran; this experiment was cancelled after one flight due to funding problems. After the fall of communism in 1991, the vast Baikonur facility fell on hard times, but it eventually revived in step with new trends in space. The Baikonur launch pads found new purpose with the advent of the International Space Station (ISS), assembly of which began in orbit in 1998. At this juncture, Russia has employed not only the Soyuz rockets, but the Proton, Zenit, and Tsyklon types—all mobilized for commercial purposes, including launching satellites for other nations and private corporations. At the turn of the 21st century, the Russian space program continued to respond to the growing demand by nations and telecommunications companies for new satellite launches.[2]

The Russian space program has taken unpredictable twists and turns in the immediate post-communist era. Most notable and controversial were experiments in what became known as "space tourism." In April

2001 they ferried the first space tourist, California millionaire Dennis Tito, to the ISS. Several more wealthy individuals have followed. These trips have been and will remain quite expensive, with the Russian space agency charging anywhere from 20 to 25 million dollars per flight. Such launches, though, have provided a welcome source of hard currency for the Russian space program.

NASA did not receive the green light for the development of a space station until the mid-1980s, a project that eventually evolved into the ISS. When finally completed, the ISS will be a remarkable accomplishment, with the interior space equivalent to a Boeing 747 jet airliner. More than 19 nations have participated in the ISS program, including space shuttle missions delivering components of the station to orbit. Convergence and global cooperation in space has become the established norm.

If manned spaceflights beyond Earth orbit were not pursued in the decades after Apollo, vast new strides were made in deep-space probes. Important recent examples include NASA's Spirit and Opportunity rovers, both of which successfully landed on Mars in 2004, using a unique airbag system to cushion their impact. The rovers explored the red planet's rocks and soil, seeking traces of the past presence of water. Sharing the rest of the world's longtime fascination with Mars, the Russians have announced a deep-space probe to the Martian moon Phobos in 2009. The unmanned spacecraft will seek to obtain samples of the moon's surface and fly them back to Earth. If successful, the three-year round-trip would be the first between Earth and the immediate vicinity of Mars.[3]

Several entrepreneurs in the United States have initiated projects for suborbital spaceflights, to be offered to the public as early as 2008. *SpaceShipOne* has attracted the most attention for its demonstrated capabilities as a non-governmental program of space travel. The design of a low-cost private space vehicle emerged from the competition for the 10-million-dollar Ansari X-Prize. In October 2004, *SpaceShipOne* made its dramatic debut. The futuristic spacecraft won the prize by carrying out two successful flights to the edge of outer space in less than two weeks. This remarkable flying machine was launched at altitude by its carrier aircraft, the *White Knight,* and then fired into the upper atmosphere by its hybrid rocket engine, fueled by a nontoxic mix of liquid nitrous oxide and rubber. The highest of the two qualifying flights, piloted by Brian Binnie, reached nearly 70 miles, the border of outer space, eclipsing the highest altitude flight of the famed X-15 rocket plane.

The interest in space travel takes more than one form. The voters in New Mexico have approved a tax increase to finance construction of America's first commercial spaceport, expected to open in 2010, near the Army's White Sands missile test facility. The spaceport is intended to provide facilities for space tourism ventures, and a number of other states—and foreign counties including Canada and Australia—are actively considering development of similar spaceports.[4]

Recent trends suggest a growing commercial interest in space. *Aviation Week & Space Technology*, a highly respected and influential aerospace trade publication, has observed: "A private spaceflight industry . . . is steadily emerging from the dusty desert hangars and closely-guarded office-park bays that incubated it, ready to leap off launch pads across the globe." More than half of the estimated 180 billion dollars in worldwide revenue generated by space-related activities such as building, launching, and operating commercial communication satellites is from privately funded activities, with the remainder from government.[5]

The number of "space-faring" nations continues to grow. The monopoly once enjoyed by the United States and Russia has ended. Seventeen members now participate in the European Space Agency (ESA), which markets itself as "Europe's gateway to space." At the turn of the 21st century no less than a dozen nations were active in satellite launches. Communication satellites alone have been a transformative force in reshaping modern life and giving expression to the idea of "globalization." The opportunity to offer commercial launch services for communications and other satellite types has been a driving force in pushing nations to develop and market their own launch complexes. China has emerged as a major player in the growing diversity of space exploration. China has been joined by two other Asian nations, Japan and India. Japan has launched its own interplanetary space probes, and India has developed satellite-delivered educational, medical, and weather services to its rural villages.[6]

Although the United States continues to be a dominant player in space, the American space program must contend with its own challenges in the coming decades. Around 2010, the space shuttle will fly its final missions, in support of the ISS and then be retired. NASA will then face a gap of at least five years without any manned spaceflight capacity until its next generation of spacecraft comes on line around 2015. During that time America will have to depend on Russia—and pay hundreds of millions of dollars—for any manned missions to the ISS.[7]

In addition, China has announced its intention to be a major space power in the 21st century, an initiative that will be long-range and aggressive. Both military and civilian programs will define the Chinese space activities in the new century. China already has demonstrated a capacity for manned spaceflight, launching its first manned space mission on October 15, 2003— Yang Liwei completed a 14-orbit mission in a Shenzhou 5 spacecraft. In October 2005, China sent two astronauts, Fei Junlong and Nie Haisheng, on a five-day orbital journey on board the Shenzhou 6 spacecraft. China has since announced ambitious plans to send its astronauts to the moon. One consequence of China's debut as a space-faring nation may be a new race with the United States to return to the moon.

In early 2004, President George W. Bush sounded a national call for Americans to return to the moon as early as 2015 with the ambitious goal of building a permanent lunar base. This same blueprint for the future also made reference to a future mission to Mars. Space visionaries greeted the call with enthusiasm. However, the Bush call for a bold future in space failed to arouse public enthusiasm, at least for the near term.

Historian Arthur M. Schlesinger, Jr., noted in 2004: "It has been almost a third of a century since human beings took a step on the moon— rather as if no intrepid mariner had bothered after 1492 to follow up on Christopher Columbus. Yet 500 years from now (if humans have not blown up the planet), the 20th century will be remembered, if at all, as the century in which man began the exploration of space."[8]

In humankind's race to explore the cosmos, much is still to come. . . .

NOTES

PROLOGUE

1. Walter Dornberger, *V-2*, New York: Viking Press, 1958, p. 8.
2. Ibid., p. 12.
3. Wernher von Braun and Frederick I. Ordway III, *Rockets' Red Glare*, New York: Anchor Press, 1976, p. 147.
4. Dornberger, p. 52.
5. Ibid., p. 17.
6. Albert Speer, *Inside the Third Reich*, New York: Colliers Books, 1981, p. 367.

CHAPTER I

1. James McGovern, *Operation Crossbow and Overcast*, New York: William Morrow & Company, Inc., 1964, pp. 99-101.
2. Von Braun and Ordway, *Rockets' Red Glare*, p. 136f.
3. Ibid., p. 140.
4. Paul Dickson, *Sputnik: The Shock of the Century*, New York: Walker & Company, 2001, p. 52.
5. William E. Burrows, *The New Ocean: The Story of the First Space Age*, New York: Random House, 1998, pp. 101-102.
6. Ibid., p. 102.
7. "Eyewitness Accounts of V-2 Bombings," BBC (audio recordings), November 1944, *Voices of World War II,* edited by Richard Lidz and Marc Goldbaum, New York: Macmillan Publishing Company, Inc., 1975.
8. T. D. Dungan, *V-2: A Combat History of the First Ballistic Missile,* Yardley, Penn.: Westholme Publishers, 2005, pp. 192-193, 201-204.
9. Charles A. Lindbergh, *Autobiography of Values,* New York: Harcourt, Inc., 1978, p. 345.
10. Ibid, p. 346.
11. Tom Bower, *The Paperclip Conspiracy, The Battle for the Spoils and Secrets of Nazi Germany,* London: Michael Joseph, 1989, p. 110.
12. Boris Chertok, *Rakety i lyudi,* Vol. 1, 2nd Edition, Moscow: Mashinostroyeniye, 1999, pp. 141-160.
13. Ibid., p. 92.
14. Boris Chertok, *Rockets and People* [English translation], Edited by Asif A. Siddiqi, Washington, D.C.: National Aeronautics and Space Administration, 2005, p. 213; *Dorogi v kosmos, Vospominaniya veteranov raketno-kosmicheskoi tekhniki i kosmonavtiki,* Vol. 1, edited by Yu. A. Mozzhorii, G. S. Titov and others, Moscow: Izdatel'stvo MAI, 1992, pp. 156-158.

15. Chertok, *Rockets and People*, Vol. 1, pp. 239-241.
16. Ibid., p. 281; Asif Siddiqi, *Challenge to Apollo: The Soviet Union and the Space Race, 1945-1974*, Washington, D.C.: National Aeronautics and Space Administration, History Division, 2000, p. 28f.
17. Chertok, *Rockets and People*, Vol. 1, p. 289.
18. Steven J. Zaloga, *The Kremlin's Nuclear Sword: The Rise and Fall of Russia's Strategic Nuclear Forces, 1945-2000*, Washington, D.C.: Smithsonian Institution Press, 2002, p. 36.
19. Siddiqi, *Challenge to Apollo*, pp. 32-39.
20. Ibid., p. 30.
21. Gröttrup, Irmgard, *The Rocket Wife*, London: Andre Deutsch, 1959, as quoted on website www.russianspaceweb.com.
22. Leonid L. Kerber, *Stalin's Aviation Gulag, A Memoir of Andrei Tupolev and the Purge Era*, edited by Von Hardesty, Washington, D.C.: 1996, pp. 166-167.
23. Chertok, *Rockets and People*, Vol. 1, pp. 329-332.
24. Deborah Cadbury, *Space Race: The Epic Battle Between America and the Soviet Union for Dominion of Space*, New York: HarperCollins, 2005, p. 103; Asif Siddiqi, "The Rockets' Red Glare: Technology, Conflict, and Terror in the Soviet Union," *Technology and Culture*, Vol. 44, Number 3 (July 2003), pp. 470-501.
25. Siddiqi, *Challenge to Apollo*, p. 54.
26. Bob Ward, *Dr. Space: The Life of Wernher von Braun*, Annapolis: Naval Institute Press, 2005, p. 70.
27. Walter A. McDougall, *... The Heavens and the Earth: A Political History of the Space Age*, Baltimore: Johns Hopkins University Press, 1997, pp. 42-43; Ward, pp. 152-155.
28. Cadbury, pp. 105-106.
29. Daniel Lang, *The New Yorker*, Vol. 24 (July 24, 1948), pp. 40-46.
30. Ernst H. Krause, "High Altitude Research with V-2 Rockets," Proceedings of the American Philosophical Society, Vol. 91, No. 5 (December 3, 1947), pp. 430-446
31. Burrows, p. 135.
32. "A Brief History of Animals in Space," National Aeronautics and Space Administration, see http://history.nasa.gov/animals.html.
33. Steven J. Zaloga, *V-2 Ballistic Missile, 1942-52*, New York: Osprey Publishing, 2003, p. 39.
34. Ward, p. 74.
35. Ibid., p. 75.
36. Frederick L. Ordway III and Mitchell R. Sharpe, *The Rocket Team*, Burlington, Canada: Apogee Books, 2003, p. 249.

CHAPTER 2

1. John Lewis Gaddis, *The Cold War: A New History*, New York: Penguin Press, 2005, p. 35.
2. Willy Ley, *The Conquest of Space*, New York: Viking Press, 1949.
3. *The Saturday Review*, December 24, 1949, p. 9.
4. Ibid.
5. *Collier's*, October 18, 1952, p. 74.
6. Ibid.
7. Ibid.
8. Ibid., pp. 52-56.
9. *Collier's*, April 30, 1954, pp. 24-25.
10. Mike Wright, "The Disney-Von Braun Collaboration and its Influence on Space Exploration," MSFC History Office, 1993, http://history.msfc.nasa.gov/vonbraun/disney_article.html; *Collier's*, April 30, 1954 pp. 24-25.
11. Dwayne Day, "Science Fiction Literature," U.S. Centennial of Flight Commission, http://www.centennialofflight.gov/essay/Social/space_lit/SH10.html.
12. Cargill Hall, "The Eisenhower Administration and the Cold War: Framing Astronautics to Serve National Security," Prologue: *Quarterly Journal of the National Archives*, Vol. 27, No.1 (1995), pp. 59-72.

13. Burrows, pp. 156-157.
14. Hall, pp. 59-72.
15. Ibid.; Burrows, p.158.
16. Walter A. McDougall, ... *The Heavens and the Earth,* Baltimore: Johns Hopkins University Press, 1997, p. 116; Burrows, p. 158.
17. McDougall, p. 116.
18. McDougall, p.118.
19. Burrows, pp.127-129.
20. Burrows, p.159.
21. Ibid., p. 174.
22. T. A. Heppenheimer, *Countdown: A History of Space Flight,* New York: John Wiley & Sons, 1997, p. 98.
23. Zaloga, *The Kremlin's Nuclear Sword,* pp. 13-15, 23-24; see also Boris Chertok, *Rockets and People, Volume 2: Creating a Rocket Industry* (English translation of Rakety i lyudi), edited by Asif Siddiqi, Washington, D.C.: National Aeronautics and Space Administration, History Division, 2006, pp. 230-231.
24. Ibid.
25. McDougall, p. 98.
26. Heppenheimer, p. 73.
27. Ibid., pp. 72-73.
28. McDougall, p. 106.
29. Ward, p. 80.
30. Ordway and Sharpe, pp. 253-254.
31. See website http://www.astronautix.com/lvs/redstone.html.
32. Siddiqi, *Challenge to Apollo,* pp. 90-91.
33. Zaloga, p. 38.
34. Siddiqi, *Challenge to Apollo,* p. 111.
35. Zaloga, p. 39.
36. Siddiqi, *Challenge to Apollo,* p. 112.
37. Ibid.
38. Chertok, *Rockets and People,* Vol. 2, pp. 284-285.
39. Ibid.; "Khrushchev Remembers," quoted in Siddiqi, *Challenge to Apollo,* p. 117.
40. Cadbury, pp. 138-139.
41. Ibid; 48. Siddiqi, *Challenge to Apollo,* p. 129.
42. Cadbury, p. 139.
43. Siddiqi, *Challenge to Apollo,* p. 130.
44. Ibid., pp.130-131.

45. Cadbury, p. 140.
46. Siddiqi, *Challenge to Apollo,* pp. 130-132.
47. Ibid., p.133.
48. McDougall, p. 101.
49. Ibid., p. 102.
50. Matt Bille and Erika Lishock, *The First Space Race: Launching the World's First Satellites,* College Station: Texas A & M University Press, 2004, p. 29.
51. McDougall, p. 109.
52. Ibid., p. 110.
53. Siddiqi, *Challenge to Apollo,* pp. 84-85.
54. Ibid., p. 86.
55. NSC 5520, quoted in Burrows, p. 166.
56. NSC 5520 and NSC 5522, quoted in Burrows, pp. 166-168.
57. Bille and Lishock, pp. 74-75.
58. "James A. Van Allen, Discoverer of Earth-Circling Radiation Belts, Is Dead at 91," *The New York Times,* (August 10, 2006), p. C14.
59. "Reach into Space," *Time,* May 4, 1959, see http://www.time.com/time/archive.
60. Burrows, pp. 168-169.
61. Quoted in Burrows, p. 169.
62. Burrows, p. 169; and Heppenheimer, *Countdown,* p. 93.
63. Ward, p. 94.
64. Ibid.; Bille and Lishock, pp. 44-45.
65. Bille and Lishock, p. 48; Ward, p. 95; Burrows, p. 170.
66. Bille and Lishock, p. 51; Ward, p. 95.
67. Burrows, pp. 170-171.
68. Bille and Lishock, pp. 76-77.
69. McDougall, p. 121.; Bille and Lishock, p. 52.
70. Ibid., p. 77; Asif Siddiqi, "Korolev, Sputnik, and the International Geophysical Year," see http://www.hq.nasa.gov/office/pao/History/sputnik/siddiqi.html.
71. Ibid., Bille and Lishock, p.77.
72. Ibid., p. 80; McDougall, p. 123.
73. Burrows, p. 171.
74. McDougall, p. 123-124.
75. Bille and Lishock, p. 118.
76. Burrows, p. 172.
77. Bille and Lishock, pp. 91, 94-95, 98.

78. Ibid., p. 99.

79. Ibid., p. 116.

80. Bille and Lishock, pp. 116-117.

81. Ibid., p. 117.

82. T. A. Heppenheimer, "How America Chose Not to Beat Sputnik into Space," *Invention and Technology,* Winter 2004, p. 44; *Countdown: A History of Spaceflight,* p. 100; Ward, p. 98.

83. Boris Chertok, *Rockets and People,* Vol. 2, pp. 419-433; James Harford, *Korolev: How One Man Masterminded the Soviet Drive to Beat the Americans to the Moon,* New York: John Wiley & Sons, 1999, p. 126. (hereafter Korolev).

84. Sidiqqi, *Challenge to Apollo,* pp. 164-166.

85. Willy Ley, *The Conquest of Space,* New York: Viking Press, 1949.

86. Robert Godwin, ed., *Mars: The NASA Reports,* Wheaton, Il.: Apogee Books, 2000.

CHAPTER 3

1. Chertok, *Rockets and People,* Vol. 2, p. 383.

2. Quoted in Dickson, p. 105.

3. Quoted in James Harford, "Korolev's Triple Play: Sputniks 1, 2, and 3," in http://history.nasa.gov/sputnik/harford. html., p. 6.

4. Ibid., p. 7.

5. Chertok, *Rockets and People,* Vol. 2, p. 385.

6. See "Russian Launch Vehicles," http:// www.century-of-flight.freeola.com.

7. Siddiqi, *Challenge to Apollo,* pp. 166-167.

8. Siddiqi, p. 168; Dickson, pp. 106-107; Ari Sternfeld, *Soviet Writings on Earth Satellites and Space Travel,* New York: Citadel Press, 1985, p. 63.

9. See William Walter, *The Space Age,* pp. 79-80; "Soviet Embassy Guests Hear of Satellite From an American as Russians Beam," *New York Times,* October 5, 1957, p. 3.

10. Harford, "Korolev's Triple Play: Sputniks 1, 2, and 3," pp. 9-13.

11. Ibid.

12. See *Time,* October 14, 1957; Sergei N. Khrushchev, *Nikita Khrushchev and the Creation of a Superpower,* University Park: Pennsylvania State University Press, 2000, pp. 259-262.

13. Sternfeld, pp. 29-32; 63-79.

14. Chertok, *Rockets and People,* Vol. 2, p. 388.

15. Ibid, pp. 388-389; Harford, "Korolev's Triple Play: Sputniks 1, 2, and 3," pp. 10-13.

16. Burrows, pp. 198-199.

17. Martin Caidin, *Vanguard! The Story of the First Man-Made Satellite,* New York: Dutton & Company, Inc., 1957, p. 13.

18. Burrows, p. 190.

19. Burrows, pp. 191-192; see http://www. loc.gov/exhibits/mead/oneworld-learn. html.

20. Quoted in the "Sputnik Triumph" in www.century-of-flight.freeola.com.

21. James Hansen, *First Man, The Life of Neil Armstrong,* New York: Simon & Schuster, 2005, pp. 168-169.

22. Editorial reprinted in *University Daily Kansan,* Wednesday, October 16, 1957, p. 2.

23. *University Daily Kansan,* Monday, November 18, 1957, p. 8.

24. Quoted in "Ike Doesn't Know about Americans," *University Daily Kansan,* October 28, 1957, p. 9.

25. Harry S. Truman, Speech, Jefferson-Jackson Dinner, Los Angeles, November 1, 1957, General Files for 1957, Truman Presidential Library and Archive.

26. Walter, *The Space Age,* p. 84; Burrows, pp. 204-205.

27. Ibid., p. 188; Ward, p. 111.

28. Ward, pp. 113-114.

29. Burrows, pp. 206-207.

30. "James A. Van Allen, Discoverer of Earth-Circling Radiation Belts, Is Dead at 91," *New York Times,* August 10, 2006.

31. Burrows, pp. 207-208.

32. Burrows, p. 209; Ward, p. 115.

33. Burrows, pp. 207-209; Howard E. McCurdy, *Inside NASA: High Technology and Organizational Change in the U.S.*

Space Program, Baltimore: Johns Hopkins University Press, 1993, pp. 16-17.

34. Ward, pp. 115-116; Burrows, p. 209.

35. Ibid., p. 228.

36. See "Red Files: Secret Soviet Moon Mission—Dr. Sergei Khrushchev" www.PBS.org/redfiles.

37. Chertok, *Rakety i lyudi, Fili, Podlipki, Tyuratam,* Book Two, Moscow: Mashinostroyeniye, 1999, pp. 251-257.

38. Siddiqi, *Challenge to Apollo,* p. 527f.

39. "Exploration of Space," http://www.century-of-flight.freeola.com; Heppenheimer, *Countdown,* p.173.

40. Ibid., p. 156.

41. Burrows, pp. 214-215.

42. Ibid., pp. 216-218.

43. Dickson, p. 227; McDougall, p. 160.

44. Burrows, pp. 229, 232-233.

45. Burrows, p. 166; "Atlas D," http://www.astronautix.com/lvs/atlasd.htm.

CHAPTER 4

1. Howard E. McCurdy, *Space and the American Imagination,* Washington, D.C.: Smithsonian Institution Press, 1997, p. 60.

2. Tom D. Crouch, *Aiming for the Stars: The Dreamers and Doers of the Space Age,* Washington, D.C.: Smithsonian Institution Press, 1999, p. 154.

3. Michael Gorn, *NASA: The Complete Illustrated History,* New York: Merrell, 2005, pp. 83-93.

4. Crouch, p. 155.

5. Ibid., p. 156; Burrows, 287.

6. Martin Caidin, *Man into Space,* New York: Pyramid Books, 1961, p. 158.

7. Burrows, pp. 284-289.

8. Tom Wolfe, *The Right Stuff,* New York: Farrar Straus and Giroux, 1979, p. 100.

9. Burrows, pp. 290-292.

10. Hugh Sidey, *John F. Kennedy, President,* Greenwich: Fawcett Publications, Inc., p. 113.

11. McDougall, pp. 221-222.

12. Burrows, pp. 268, 305; see also http://www.earth.nasa.gov/history/tiros.html.

13. "Chronology of Two-Year Dispute on 'Missile Gap,'" *The New York Times,* Feb. 9, 1961, p. 4.

14. John Rhea, Editor, *Roads to Space: An Oral History of the Soviet Space Program,* New York: McGraw-Hill Companies, 1995, p. 375.

15. See Dwayne A. Day, "Of Myths and Missiles: The Truth about John F. Kennedy and the Missile Gap," http://www.thespacereview.com, January 3, 2006.

16. Ibid.

17. Ibid.

18. McDougall, p. 222.

19. Ibid., pp. 311-312.

20. Ibid., p. 311; Gorn, p. 99.

21. Burrows, p. 320; Gorn, p. 100.

22. Jack Valenti, "Thank the Lord for the Good Men," *Houston Post,* March 31, 1962, a news clipping in James Webb Papers, Harry S. Truman Library and Archive.

23. Burrows, pp. 265-266; Gorn, p. 107.

24. Roger D. Launius, *Apollo: A Retrospective Analysis, The Kennedy Perspective on Space,* p. 1, in http://www.hq.nasa.gov/office/pai/History/Apollomon/Apollo.html.

25. Sidey, p. 62.

26. Launius, *Apollo,* p. 1.

27. Gorn, p. 82; Bille and Lishock, pp. 152-153.

28. Burrows, p. 294.

29. Wolfe, *The Right Stuff,* p. 58.

30. Burrows, 307; Chertok, *Rakety i lyudi,* Vol. 2., Moscow: Mashinostroyeniye, 1999, 430-431.

31. Burrows, pp. 308-309; Siddiqi, *Challenge to Apollo,* pp. 98, 256-258.

32. Chertok, *Rakety i lyudi,* Vol 2., pp. 426-427.

33. N. P. Kamanin, *Skrytiy kosmos, Kniga pervaya,* 1960-1962, Moscow: Infortekst, 1995, p. 99.

34. Siddiqi, *Challenge to Apollo,* pp. 261-262.

35. Harford, *Korolev,* pp. 171-176; Siddiqi, *Challenge to Apollo,* pp. 274-282.

36. Siddiqi, *Challenge to Apollo,* p. 277.
37. Harford, *Korolev,* p. 172.
38. Siddiqi, *Challenge to Apollo,* p. 278.
39. Burrows, p. 313.
40. Ibid.; Mark Gallai, *S Chelovekom na bortu: povesti,* Moscow: Sovetskiy Pisatel', 1985, pp. 122-126.
41. Siddiqi, *Challenge to Apollo,* p. 280.
42. Harford, *Korolev,* p. 174.
43. Siddiqi, *Challenge to Apollo,* p. 281.
44. Burrows, p. 315; Chertok, *Rakety i lyudi,* Vol. 2, p. 421; Gallai, pp. 126-131.
45. Sidey, pp. 109-111.
46. Roger Launius, *Apollo,* p. 3.
47. Sidey, pp. 118-119; Dwayne Day, "Pay no attention to the man with the notebook: Hugh Sidey and the Apollo Decision," December 5, 2005 in http://www.thespacereview.com/article/511/1.
48. Sidey, p.120.
49. Day, "Pay no attention to the man with the notebook: Hugh Sidey and the Apollo Decision."
50. Launius, *Apollo,* pp. 3-4.
51. Dwayne Day, "A new space council?," June 21, 2004, http://www.thespacereview.com/article/163/1.
52. Burrows, p. 189; Heppenheimer, *Countdown,* p. 126.
53. See http://history.nasa.gov/Apollomon/docs.htm.
54. Launius, *Apollo,* p. 4.
55. Burrows, pp. 266-267; McDougall, p. 225; and see "Chronology of Major Events in Manned Spaceflight" in http://www.hq.nasa.gov/pao/History/SP-4214/app4/html.
56. Excerpted from http://history.nasa,gov/Apollomon/docs.html.
57. McNamara and Webb memo to Johnson, quoted in Burrows, pp. 328-329.
58. Burrows, pp. 328-329.
59. "Special Message to the Congress on Urgent National Needs," John F. Kennedy's remarks to a Joint Session of Congress, dated May 25, 1961.
60. "An Historic Meeting at the White House on Human Spaceflight," http://www.history.nasa.gov/JFK-Webbconv/.
61. Sidey, pp. 152-154.
62. Wolfe, *The Right Stuff,* p. 211.
63. Robert C. Seamans, Jr., *Project Apollo, The Tough Decisions,* Monographs in Aerospace History, No. 37-SP-2005-4537, National Aeronautics and Space Administration, 2005, p. 18.
64. Lloyd S. Swenson, Jr., James M. Grimwood, and Charles C. Alexander, "This New Ocean: A History of Project Mercury," Shepard's Ride, p. 1, in http://www.hq.nasa.gov/office/pao/History/SP-4201/toc.htm.
65. Seamans, p. 30; "This New Ocean: A History of Project Mercury," Liberty Bell Tolls, pp. 1-3.
66. Siddiqi, *Challenge to Apollo,* pp. 290-291.
67. Ibid., p. 292.
68. Ibid., p. 260; 291.
69. Burrows, pp. 338-340.
70. Siddiqi, p. 294.
71. Ibid.
72. Ibid., pp. 294-295.
73. Wernher von Braun, *Across the Space Frontier,* New York: Viking, 1952.

CHAPTER 5

1. Svetlana Boym, *Kosmos: A Portrait of the Russian Space Age,* Princeton: Princeton Architectural Press, 2002, frontispiece.
2. Ibid, p. 90.
3. David West Reynolds, *Kennedy Space Center: Gateway to Space,* Buffalo: Firefly Books, 2006.
4. Charles Lindbergh, *Autobiography of Values,* New York: Harcourt, Inc., 1976, p. 343.
5. Ibid., p. 344.
6. Peter P. Wegener, *The Peenemünde Wind Tunnels, A Memoir,* New Haven: Yale University Press, 1996, pp. 16-17.
7. David Scott and Alexei Leonov, *Two Sides of the Moon: Our Story of the Cold War Space Race,* New York: St. Martin's Press, 2004, pp. 96-97; Siddiqi, *Challenge to Apollo,* p. 135.

8. Burrows, p. 164.

9. Boym, p. 90.

10. See *The Bulletin of the Atomic Scientists,* 54, No. 5 (September 1998), pp. 44-50.

11. Walter Duranty, *USSR, The Story of Soviet Russia,* New York: J.B. Lippincott Company, 1944, p. 139.

12. Boym, *Kosmos,* p. 91.

13. Siddiqi, *Challenge to Apollo,* p. 134.

14. David Scott and Alexei Leonov, *Two Sides of the Moon, Our Story of the Cold War Space Race,* New York: St. Martin's Press, 2004, p. 97.

15. Siddiqi, *Challenge to Apollo,* pp. 135-136.

16. Chertok, *Rockets and People,* Vol. 2, p. 314.

17. Scott and Leonov, pp. 96-97.

18. Chertok, *Rockets and People,* Vol. 2, p. 315.

19. Ibid., p. 316; see www.astronautix.com/sites/Baikonur.html.

20. Chertok, *Rockets and People,* Vol. 2, p. 327.

21. Zaloga, *The Kremlin's Nuclear Sword,* p. 47.

22. Dwayne A. Day, "The Secret at Complex J," *Air Force Magazine,* July 2004, pp. 72-75; Dino A. Brugioni, "The Tyuratam Enigma," *Air Force Magazine,* July 1984, pp. 108-109.

23. Siddiqi, *Challenge to Apollo,* p. 258; Scott and Leonov, pp. 55-56; James Oberg, "Disaster at the Cosmodrome," *Air & Space/Smithsonian* magazine, December 1990, pp. 74-77; Chertok, *Rockets and People,* Vol. 2, p. 598.

24. Siddiqi, *Challenge to Apollo,* p. 266.

25. Heppenheimer, *Countdown,* pp. 330-331.

26. Ward, pp. 70-71; Ordway and Sharpe, *The Rocket Team,* p. 243.

27. Martin Caidin, *Spaceport U.S.A: The Story of Cape Canaveral and the Air Force Missile Test Center,* New York: E.P. Dutton & Company, 1959, pp. 68-69; David West Reynolds, *Kennedy Space Center,* pp. 62-63.

28. "The Making of the Cape," Moonport: A History of Apollo Launch Facilities and Operations, in http://www.hq.nasa.gov/office/pao/History/SP-4204/ch1-3.html; Ward, pp. 80-81.

29. Heppenheimer, *Countdown,* p. 116.

30. Reynolds, pp. 63-65; Caidin, *Spaceport U.S.A.,* p. 105.

31. Caidin, p. 106; Reynolds, p. 63-65.

32. Joel Powell, *Go For Launch: An Illustrated History of Cape Canaveral,* Burlington (Canada): Apogee Books, 2006, pp. 50-53; Reynolds, pp. 71-72.

33. Heppenheimer, *Countdown,* p.120.

34. Caidin, *Spaceport U.S.A.,* pp. 29-35.

35. Reynolds, p. 33.

36. Caidin, *Spaceport U.S.A.,* pp. 37-63.

37. Ibid.

38. Powell, p.127.

39. Gorn, p. 107.

40. "Life in Missileland," *Time,* July 15, 1957, in http://www.time.com/time.

41. "The Future of NASA," *Time* (August 10, 1970) and "Apollo: Where Is Its Poetry?" *Time* (August 9, 1971) in http://www.time.com/time.

42. "A New Look at the Cape," *Time* (March 26, 1965) in http://www.time.com/time.

43. Gorn, pp. 103, 107; Reynolds, *Kennedy Space Center,* p. 40.

44. Ibid., pp. 26-27.

45. Wernher von Braun, *Across the Space Frontier,* New York: Viking, 1952.

CHAPTER 6

1. "40th anniversary of Glenn's Orbital Flight," in www.npr.org/programs/atc/features/2002.

2. "First Orbit—25th Anniversary," Lecture by John Glenn, National Air and Space Museum, February 25, 1987.

3. Burroughs, p. 342; Wolfe, *The Right Stuff,* pp. 267-268.

4. Burroughs, p. 341; Hansen, p. 211.

5. See Glenn, "First Orbit—25th Anniversary."

6. Burrows, pp. 340-342.

7. Quoted in *Men in Space*, p. 45.

8. See *Time*, March 2, 1962; Scott and Leonov, pp. 76-77; Siddiqi, *Challenge to Apollo*, pp. 356-361.

9. Ibid., p. 358.

10. Ibid., p. 360.

11. Robert C. Seamans, Jr., *Project Apollo, The Tough Decisions*, Washington, D.C.: NASA History Division, 2005, p. 42.

12. Siddiqi, *Challenge to Apollo*, p. 371.

13. *Skrytyy Kosmos, Book One*, Moscow: Infortekst, 1995, p. 182.

14. Siddiqi, *Challenge to Apollo*, p. 362.

15. Scott and Leonov, p. 77.

16. Ibid.

17. Siddiqi, *Challenge to Apollo*, pp. 372-373.

18. "Salyut, Soviet Steps Toward Permanent Human Presence in Space, A Technical Memorandum," Washington, D.C.: Office of Technology Assessment, United States Congress, December 1983, pp. 10-11.

19. Scott and Leonov, p. 77.

20. Ibid., pp. 53, 79; Khrushchev, pp. 686-687, 691-692; see also, Yaroslav Golovanov and Sergei Korolev, *The Apprenticeship of a Space Pioneer*, translated by M. M. Samokhvalov and H. C. Creighton, Moscow: Mir Publishers, 1975, and *Korolev: Fakty i mify*, Moscow: Nauka, 1994.

21. Scott and Leonov, p. 78.

22. Ibid; Memoir of Evgeniy V. Shabrov in *Dorogi v kosmos*, edited by Yu. A. Moszhorin and others, Moscow: Izdatel'stvo MAI, 1992, p. 179.

23. Ibid.

24. Scott and Leonov, p. 99.

25. Ibid. p. 100.

26. Ibid., 105.

27. Scott and Leonov, p. 109.

28. Ibid.

29. Ibid.

30. Ibid.

31. Ibid.

32. Launius, *Apollo: A Retrospective Analysis*, pp. 14-15; Gorn, pp. 100, 103.

33. Ibid., pp.103, 108.

34. Astronaut Biographical Data, inhttp://www.jsc.nasa.gov/Bios/htmlbios/white-eh.html.

35. Burrows, pp. 358-359; Barton C. Hacker and James M. Grimwood, "Four Days and a 'Walk,'" On The Shoulders of Titans: A History of Project Gemini, pp. 1-3 see http://history.nasa.gov/SP-4203/toc.htm.

36. Hacker and Grimwood, pp. 1-2.

37. Seamans, pp. 62-63.

38. Heppenheimer, *Countdown*, p. 222; see also http://history.nasa.gov/ap11ann/astrobios.htm.

39. Heppenheimer, *Countdown*, pp. 222-223.

40. Reynolds, pp. 109-110; Andrew Chaikin, *A Man on the Moon: The Voyages of the Apollo Astronauts*, New York: Penguin Books, 1995, pp. 141-143.

41. Seamans, p. 67; see http://history.nasa.gov/ap11ann/astrobios.htm.

42. Gorn, p. 116.

43. Burrows, pp. 362-363.

44. Burrows, pp. 364-369, 388-390; Asif Siddiqi, "Robotic U.S. Missions to the Moon," pp. 1-2, U. S. Centennial of Flight Commission, http://centennialofflight.gov/essay_cat/11.htm

45. Ibid.

46. Burrows, pp. 391-394; "Robotic U.S. Missions to the Moon," p. 2.

47. "Soviet Lunar Missions," http://nssdc.gsfc.nasa.gov/planetary/lunar/lunarussr.html; Reynolds, Apollo, p. 59.

48. Burrows, p.354 and Reynolds, Apollo, p. 59.

49. "Robotic U.S. Missions to the Moon," p. 2.

50. Reynolds, Apollo, p. 59.

51. Burrows, p. 395, 401-402.

52. "Soviet Lunar Missions," ibid.

53. Asif Siddiqi, "Exploration of the Inner Planets," pp. 1-2, U.S. Centennial of Flight Commission, http://www.centennialofflight.gov/essay_cat/11.htm.

54. "Exploration of the Inner Planets," pp. 2-3; Burrows, pp. 460-461.

55. Dwayne Day, "Exploration of the Outer Planets," p. 1, U.S. Centennial of Flight Commission, http://www.centennialofflight.gov/essay_cat/11.htm.

56. Dwayne Day, "Commercial Communications Satellites," pp 1-2, U.S. Centennial of Flight Commission," http://www.centennialofflight,gov/essay_cat/11.htm.

CHAPTER 7

1. Launius, *Apollo: A Retrospective Analysis,* p. 7.
2. Apollo: Expeditions to the Moon, "A Perspective on Apollo," in http://www.hq.nasa.gov/office/pao/History/SP-350/toc.html.
3. Launius, *Apollo: A Retrospective Analysis,* p. 8.
4. Burrows, p. 334.
5. Gorn, p. 112; Burrows, p. 370.
6. Ibid., p. 370.
7. Gorn, p. 112; Burrows, p. 370.
8. Burrows, pp. 370-371; Reynolds, *Apollo,* p. 81.
9. Burrows, p. 371.
10. Launius, *Apollo: A Retrospective Analysis,* pp. 8-9; Gorn, p. 149.
11. Gorn, p. 107.
12. Arnold S. Levine, "Managing NASA in the Apollo Era," Introduction, p. 2, http://history.nasa.gov/SP-4102/intro.htm; and Burrows, p. 372.
13. Burrows, pp. 405-406.
14. Apollo: Expeditions to the Moon, "An All-Up Test for the First Flight," http://history.nasa.gov/SP-350/ch-3-4.html.
15. Gorn, pp. 121, 123.
16. Chaikin, *A Man on the Moon,* pp. 11-12; Launius, *Apollo: A Retrospective Analysis,* p. 16; Kelly A. Giblin, "Fire in the Cockpit," *Invention and Technology Magazine,* Vol. 13, Issue 4 (Spring 1998), pp. 1-7, at www.AmericanHeritage.com.
17. Chaikin, *A Man on the Moon,* pp. 17-18; http://www.hq.NASA.gov/office/pao/history/Apollo204/chariot.html.
18. Burrows, pp. 410-411; Chaikin, *A Man on the Moon,* p. 24; www.hq.nasa.gov/office/pao/History/Biographies/Webb.
19. "Quiet Victims of Soviet Space Exploration," http://english/pravda.ru/main/2003/26/43726.html; John Charles, "Could the CIA Have Prevented the Apollo 1 Fire?," *The Space Review;* see www.thespacereview.com/article/797/1; "The Oxygen Question," *Time,* February 10, 1967 in www.time.com.
20. Scott and Leonov, pp. 192-193.
21. Ibid., pp. 197-198; Sidiqqi, *Challenge to Apollo,* pp. 576-590; Chertok, *Rakety i lyudi, goryachiye dni kholodnoi voiny,* Book 3, Moscow: Mashinostroyeniye, 1999, pp. 445-458.
22. Roger E. Bilstein, "Aerospace Alphabet: ABMA, ARPA, MSFC," *Stages to Saturn: A Technological History of the Apollo/Saturn Launch Vehicles,* pp. 1-2, http://history.nasa.gov/SP4206/ch2.htm.
23. Roger Bilstein, "Missions, Modes, and Manufacturing," *Stages to Saturn: A Technological History of the Apollo/Saturn Launch Vehicles,* p. 17, http://history.nasa.gov/SP-4206/sp4206.htm.
24. Launius, *Apollo: A Retrospective Analysis,* p.16.
25. Burrows, pp.261, 265; Cadbury, pp.187-188.
26. David West Reynolds, *Apollo,* p. 264.
27. Burrows, p. 373.
28. Reynolds, *Apollo,* p. 82-83.
29. Brian Berger, "Agency Spotlight: Weathering the Storms at NASA," http://space.com/spacenews/archive06/Stennis_051506.html.
30. Andrew Chaikin, *A Man on the Moon,* pp. 154-156.
31. Ibid.
32. Seamans, pp. 52-53.
33. *National Intelligence Estimate,* Number 11-1-65, submitted by the Director of Central Intelligence, concurred in by the United States Intelligence Board, dated 27 January, 1965, p.1.
34. Ibid., p. 12.

35. Chertok, *Rakety i lyudi, Goryachiye dni kholodnoi voiny,* Book 3, pp. 360-379.

36. Quoted in James Harford, "Korolev, Mastermind of the Soviet Space Program," *Cosmos,* Journal of the Cosmos Club of Washington, D.C., Washington, D.C.: Vol. 8 (1998), p. 59.

37. Scott and Leonov, p. 144; Chertok, *Rakety i lyudi, Goryachiye dni kholodnoi voiny,* Book 3, pp. 360-379; Sidiqqi, *Challenge to Apollo,* pp. 513-515.

38. Ronald D. Humble, *The Soviet Space Programme,* London: Routledge, 1989, p. 10; Sidiqqi, *Challenge to Apollo,* pp. 517-521.

39. Ibid., pp. 395-396.

40. Ibid., pp. 483-511; Harford, *Korolev,* pp. 258-268.

41. Burrows, p. 419.

42. Scott and Leonov, pp. 239-240; Cadbury, pp. 323-324; Sidiqqi, *Challenge to Apollo,* pp. 688-691.

43. Cadbury, p. 338.

44. "Project: Apollo 8" (Press Kit), National Aeronautics and Space Administration, Washington, D.C., December 15, 1968, pp. 1-2.

45. Michael Collins, *Carrying the Fire: An Astronaut's Journeys,* New York: Farrar, Straus and Giroux, 1974, pp. 296-297.

46. Collins, p. 296.

47. Reynolds, *Apollo,* pp. 72-73; Chaikin, p. 59.

48. Reynolds, *Apollo,* pp. 73-74; Chaikin, pp. 76-77.

49. Reynolds, *Apollo,* pp. 80; Burrows, p. 417.

50. Chaikin, pp. 60-67.

51. Crouch, p. 218.

52. Chaikin, pp. 68, 80.

53. "Crawler-Transporter" in http://science.ksc.nasa.gov/facilities/crawler.html.

54. Reynolds, *Apollo,* pp. 85, 92, 96.

55. Chaikin, p. 80-81.

56. Dwayne A. Day, "Saturn's fury: Effects of a Saturn 5 launch pad explosion," *The Space Review,* www.thespacereview.com/article/591/1; Reynolds, *Apollo,* p. 85.

57. Chaikin, pp. 86-88.

58. "Project: Apollo 8," p. 4; Chaikin, p. 89.

59. Collins, p. 379.

60. Burrows, pp. 375-376.

61. Collins, p. 278.

62. Ibid. p. 279.

63. "The Lunar Module Computer," Australian Broadcasting Corporation, http://abc.net/au/science/moon/computer.htm and James E. Tomayko, Computers in Spaceflight: The NASA Experience, "The need for an on-board computer," Chapter Two, http://history.nasa.gov/computers/Ch2-1.html.

64. Chaikin, p. 72.

65. Cyrus Farivar, "In Computer Years, Apollo Replica's an Antique," *New York Times,* Feb. 10, 2005.

66. Chaikin, p. 73.

67. Collins, p. 310.

68. Chaikin, p. 107.

69. Ibid., pp.121-122.

70. Ibid., pp. 112-113, 119; *Reynolds, Apollo,* pp. 110-111; Burrows, p. 420.

71. Burrows, pp. 419-421.

72. Ibid., p. 374.

73. Chaikin, p.136.

74. Reynolds, *Apollo,* pp. 124-125.

75. Ibid., pp.125, 128; Chaikin, p. 141.

76. Ibid., pp. 152-156.

77. Ibid., pp. 157-159; Reynolds, *Apollo,* p. 128.

78. Chaikin, Ibid.; Reynolds, *Apollo,* p. 128.

79. Collins, p. 296.

80. "Apollo 11 Landing Mission," Press Kit, National Aeronautics and Space Administration, Washington, D.C., dated July 6, 1969, pp. 1 and 3.

81. Collins, p. 323.

82. Ibid., p. 379.

83. "Project: Apollo 11," Press Kit, National Aeronautics and Space Administration, Washington, D.C., pp. 4-6.

84. Collins, pp. 387-388.

85. Collins, p. 400; Chaikin, pp.192-200.

EPILOGUE

1. Craig Covault, "Blame It on Nixon," *Aviation Week & Space Technology*, March 19/26, 2007, pp. 82-85.

2. Nick Allen, "Russia's Space Odyssey (Miracles on a Shoestring)," *Russian Life*, September 2003 (online).

3. Craig Covault, "Russian Exploration," *Aviation Week & Space Technology*, pp. 154-155, July 17, 2006.

4. Marc Kaufman, "N.M. County Passes Tax to Fund Spaceport," *The Washington Post*, p. A3, April 6, 2007; Ed Regis, "Field of Dreams," *Air & Space*, pp. 66-67, April/May 2007.

5. Frank Morring, Jr., "'New Space,'" *Aviation Week & Space Technology*, pp. 44-45, December 11, 2006.

6. Frank Morring, Jr., "Not Just for Superpowers," *Aviation Week & Space Technology*, pp. 71-75, March 19/26, 2007.

7. Mike Schneider, "NASA Worries about 5-year gap after shuttle retires," http://usatoday.com/tech/science/space/2007-03-30-spaceflight-gap_N.htm.

8. Arthur Schlesinger, Jr., "State of the 'Vision Thing,'" *Los Angeles Times*, January 21, 2004.

INDEX

ILLUSTRATION CREDITS

Maps by National Geographic Book Publishing Group and Justin Morrill, The M Factory.

ii, Otis Imboden/NGS Archives; vi, Conway Stuart/Sipa; xx, Rykoff Collection/CORBIS; xxviii, Ullstein Bild/The Granger Collection, NY; 8, dpa/dpa/CORBIS; 36, NASA; 69, NASA; 70, Bill Bridges/Time Life Pictures/Getty Images; 96, NRO; 98, AFP/Getty Images; 133, NASA, courtesy J.L. Pickering; 136, Doug Martin; 163, NASA, courtesy Kipp Teague; 166, NASA, courtesy Kipp Teague; 199, NASA; 202, NASA; 240, NASA, courtesy Kipp Teague.

The following photos are numbered in order of appearance:
GALLERY 1: (1), Russian State Archive of Scientific and Technical Documents; (2), CORBIS; (3), Walter Frentz/Ullstein; (4), Schenectady Museum/Hall of Electrical History Foundation/CORBIS; (5), ITAR-TASS; (6), Ullstein Bild/The Granger Collection, NY; (7), Ullstein Bild/The Granger Collection, NY; (8), NASA; (9), Sovfoto; (10), Sovfoto; (11), Bettmann/CORBIS; (12 UP), Courtesy Naum Semenovich Narovliansky; (13 LO), Courtesy Naum Semenovich Narovliansky; (14), Burt Glinn/Magnum Photos; (15), Ralph Crane/Time Life Pictures/Getty Images. **GALLERY 2:** (1), RIA Novosti; (2), Robert W. Kelley/Time Life Pictures/Getty Images; (3), James Whitmore/Time Life Pictures/Getty Images; (4), NASA; (5), Russian State Archive of Scientific & Technical Documents; (6), Ted Russell/Time Life Pictures/Getty Images; (7), Ralph Morse/Time Life Pictures/Getty Images; (8), Sergeev Aleksandr/RIA Novosti; (9), TASS/Sovfoto; (10), RIA Novosti; (11), Otis Imboden/NGS Archives; (12), Dean Conger/NGS Archives; (13), Russian State Archive of Scientific & Technical Documents; (14), Russian State Archive of Scientific & Technical Documents; (15), NASA; (16), Novosti/Topham/The Image Works; (17), Dean Conger/CORBIS; (18 UP), ITAR-TASS; (19 LO), Dean Conger/CORBIS; (20 UP), Bettmann/CORBIS; (21 LO), Bettmann/CORBIS; (22), NASA; (23), NASA, courtesy Kipp Teague; (24), NASA, courtesy Kipp Teague; (25 UP), NASA; (26 LO), NASA, courtesy Ed Hengeveld; (27), NASA, courtesy Ed Hengeveld; (28), Donald Uhrbrock/Time Life Pictures/Getty Images; (29), NASA; (30), NASA, courtesy J.J. Pickering; (31), NASA, courtesy Ed Hengeveld; (32), NASA. **GALLERY 3:** (1), NASA; (2), NASA; (3), NASA; (4), NASA; (5 UP), NASA, courtesy Kipp Teague; (7), NASA; (8), NASA; (6 lo), NASA; (9), NASA; (10), NASA, courtesy Frederic Artner; (11), NASA; (12), NASA; (13), NASA; (14), NASA, courtesy Kipp Teague; (15), NASA, courtesy Kipp Teague; (16), NASA, courtesy Kipp Teague; (17), NASA, courtesy Kipp Teague.

ABOUT THE AUTHORS

VON HARDESTY is a curator at the Smithsonian National Air and Space Museum, where he has worked since 1979. His publications include *Black Wings: Courageous Stories of African Americans in Aviation and Space, Air Force One: The Aircraft that Shaped the Modern Presidency, Lindbergh: Flight's Enigmatic Hero, Stalin's Aviation Gulag,* and *Red Phoenix: The Rise of Soviet Air Power, 1941–1945,* among many other publications. He holds a doctorate in Russian history from Ohio State University and has been a visiting fellow at the St. Antony's College, Oxford University, and the Hall Center at the University of Kansas. He has also served as a commentator on several documentaries for the History Channel and PBS, including *NOVA.* He and his wife, Patricia, live in Grottoes, Virginia.

GENE EISMAN has served as researcher and consulting editor for several books, including *Air Force One.* A former corporate public relations executive in Washington, D.C., he is a volunteer researcher at the Smithsonian National Air and Space Museum and an expert on Soviet and Russian aviation. Eisman has traveled widely in Russia and to more than 50 other countries. He and his wife, Charlene Currie, live in Bethesda, Maryland.

SERGEI N. KHRUSHCHEV, son of former Soviet Prime Minister Nikita Khrushchev, is a senior fellow at Brown University's Watson Institute for International Studies. He is the author of *Nikita Khrushchev: Crisis and Missiles* and *Nikita Khrushchev and the Creation of a Super Power,* among numerous other books and publications. An active participant in the Soviet space race as an electrical engineer, Khrushchev served from 1958-1968 as the deputy section head in charge of guidance systems for missile and space design, including moon vehicles and the "Proton," the world's largest space booster. He and his wife, Valentina Golenko, live in Cranston, Rhode Island.